# Lecture Notes in Computer Science　12023

More information about this series at http://www.springer.com/series/7408

Marco Gribaudo · Eduard Sopin ·
Irina Kochetkova (Eds.)

# Analytical and Stochastic Modelling Techniques and Applications

25th International Conference, ASMTA 2019
Moscow, Russia, October 21–25, 2019
Proceedings

Springer

*Editors*
Marco Gribaudo (iD)
Politecnico di Milano
Milan, Italy

Eduard Sopin (iD)
Peoples' Friendship University of Russia
Moscow, Russia

Irina Kochetkova (iD)
Peoples' Friendship University of Russia
Moscow, Russia

ISSN 0302-9743         ISSN 1611-3349   (electronic)
Lecture Notes in Computer Science
ISBN 978-3-030-62884-0       ISBN 978-3-030-62885-7   (eBook)
https://doi.org/10.1007/978-3-030-62885-7

LNCS Sublibrary: SL2 – Programming and Software Engineering

This Springer imprint is published by the registered company Springer Nature Switzerland AG
The registered company address is: Gewerbestrasse 11, 6330 Cham, Switzerland

# Preface

This volume contains the papers presented at the 25th International Conference on Analytical and Stochastic Modelling Techniques and Applications (ASMTA 2019), held during October 21–25, 2019, in Moscow, Russia.

Owing to the number of concurrent calls for papers in the field, the number of submissions was considerably smaller than previous years. There were 22 submissions. Each submission was reviewed by three Program Committee members. The committee decided to accept 13 papers.

This was the 25th year of ASMTA, which shows a considerable durability in a rapidly evolving field. Over the years ASMTA has been the forum for many important papers investigating the key topics of the day in the area of analytical and stochastic modeling. In this volume we are delighted to have contributions employing a diverse range of analysis techniques, including queueing theoretical results, reliability of stochastic systems, stochastic network calculus, and wide variety of applications. The range of topics within a small number of papers is impressive and demonstrates the power of stochastic analysis to tackle challenging problems in complex computer and communication systems.

We would like to take this opportunity to thank those who helped put ASMTA 2019 together, in particular Khalid Al-Begain, without whom ASMTA would not exist, and Konstantin Samouylov, who hosted the conference at RUDN University, Russia. Alexey Khokhlov was helpful in managing the conference website. We would also like to thank our colleagues in Moscow, Darya Ostrikova and Ekaterina Markova, who helped with practical arrangements and bookings, and our PhD and master's students, who acted as a local support team during the conference. Finally we would like to acknowledge the continued support of Springer in publishing the proceedings and the team at EasyChair for providing comprehensive conference support with no charge.

May 2020

Marco Gribaudo
Eduard Sopin
Irina Kochetkova

# Organization

## General Chair

Konstantin Samouylov     RUDN University, Russia

## Program Committee Chairs

Konstantin Samouylov     RUDN University, Russia
Marco Gribaudo          Politecnico di Milano, Italy

## Program Committee

Sergey Andreev          Tampere University, Finland
Jonatha Anselmi         Inria, France
Konstantin Avrachenkov   Inria, France
Christel Baier          Technical University of Dresden, Germany
Simonetta Balsamo       Università Ca' Foscari, Italy
Koen De Turck           Centrale Supélec, France
Ioannis Dimitriou       University of Patras, Greece
Antonis Economou        University of Athens, Greece
Dieter Fiems            Ghent University, Belgium
Matthew Forshaw         Newcastle University, UK
Jean-Michel Fourneau    Université de Versailles Saint-Quentin-en-Yvelines,
                            France
Yezekael Hayel          Avignon University, France
András Horváth          University of Turin, Italy
Gábor Horváth           Budapest Technical University, Hungary
Stella Kapodistria      Eindhoven University of Technology, The Netherlands
Helen Karatza           Aristotle University of Thessaloniki, Greece
William Knottenbelt      Imperial College London, UK
Lasse Leskelä           Aalto University, Finland
Daniele Manini          University of Turin, Italy
Andrea Marin            University of Venice, Italy
Yoni Nazarathy          The University of Queensland, Australia
Jose Nino-Mora          Carlos III University of Madrid, Spain
Antonio Pacheco         Institito Superior Tecnico, Portugal
Juan F. Pérez           Universidad del Rosario, Colombia
Tuan Phung-Duc          University of Tsukuba, Japan
Balakrishna J. Prabhu   LAAS-CNRS, France
Marie-Ange Remiche      University of Namur, Belgium
Anne Remke              WWU Münster, Germany
Jacques Resing          Eindhoven University of Technology, The Netherlands

| | |
|---|---|
| Marco Scarpa | University of Messina, Italy |
| Bruno Sericola | Inria, France |
| Devin Sezer | Middle East Technical University, Turkey |
| János Sztrik | University of Debrecen, Hungary |
| Miklós Telek | Budapest Technical University, Hungary |
| Nigel Thomas | Newcastle University, UK |
| Dietmar Tutsch | University of Wuppertal, Germany |
| Jean-Marc Vincent | Inria, France |
| Sabine Wittevrongel | Ghent University, Belgium |
| Verena Wolf | Saarland University, Germany |
| Katinka Wolter | Freie Universität Berlin, Germany |
| Alexander Zeifman | Vologda State University, Russia |
| María Estrella Sousa Vieira | University of Vigo, Spain |

## Additional Reviewer

Valery Naumov

# Contents

# Algorithmic Analysis of a Two-Class Multi-server Heterogeneous Queueing System with a Controllable Cross-connectivity

Dmitry Efrosinin[1,2] [ID], Irina Kochetkova[2,3(✉)] [ID], Konstantin Samouylov[2,3] [ID], and Natalia Stepanova[4] [ID]

[1] Johannes Kepler University Linz, Altenbergerstrasse 69, 4040 Linz, Austria
`dmitry.efrosinin@jku.at`
[2] Peoples' Friendship University of Russia (RUDN University),
6 Miklukho-Maklaya Street, Moscow 117198, Russian Federation
`{gudkova-ia,samuylov-ke}@rudn.ru`
[3] Institute of Informatics Problems, Federal Research Center Computer Science
and Control of the Russian Academy of Sciences,
44-2 Vavilova Street, Moscow 119333, Russian Federation
[4] V.A. Trapeznikov Institute of Control Sciences of RAS,
Profsoyuznaya Street, 65, 117997 Moscow, Russia
`natalia0410@rambler.ru`
`http://www.jku.at`, `http://eng.rudn.ru`, `http://www.ipiran.ru/english`,
`http://www.ipu.ru/en`

**Abstract.** We analyse algorithmic the queueing system with two parallel queues supplied with two heterogeneous group of servers. We assume a controllable cross-connectivity of queues with certain class of customers to different groups of servers. The system is analyzed in steady state. For a given cost structure we formulate the Markov decision problem for an optimal allocation of servers between the queues to minimize the long-run average cost per unit of time. The corresponding dynamic programming equations are derived. We develop algorithms to evaluate different performance measures including the mean busy period, the mean number of customers served in a busy period as well as the maximal queue length in a busy period. Some illustrative numerical examples are discussed.

**Keywords:** Parallel queues · Optimal allocation · Dynamic programming · Busy period · Maximal queue length

The publication has been prepared with the support of the "RUDN University Program 5-100" (recipient Konstantin Samouylov). The reported study was funded by RFBR, project numbers 18-00-01555 (18-00-01685) and 20-37-70079 (recipient Irina Kochetkova).

M. Gribaudo et al. (Eds.): ASMTA 2019, LNCS 12023, pp. 1–17, 2020.
https://doi.org/10.1007/978-3-030-62885-7_1

# 1   Introduction

The main objective of this paper is to provide algorithmic analysis in the context of the queueing system with two classes of customers. Such analysis consists of optimization part where we investigate an optimal allocation problem of servers between two parallel queues and part dedicated to evaluation of the most important performance measures.

The problem of an optimal allocation of server between the queues in multi-class systems, which is known also as a scheduling problem, has been studied extensively, see e.g. [4,5]. As it is well known, in such cases the $c\mu$-rule, called also Smith's rule or Weighted Shortest Processing Time, represents an optimal scheduling policy. In other words, the server is allocated to serve the customer from the queue with the largest index $c_i\mu_i$. The optimal control in case of non-linear costs and generalized the $c\mu$-rule was studied in [14]. Research to extend the results to incorporate switching penalties has also been conducted. In [7] and [9] the authors characterized aspects of the allocation problem for homogeneous systems. For heterogeneous systems in [6] and [8] it was proved a partial characterization of the optimal policy and provided heuristic approaches. A multi-class, multi-server queueing system with preemptive priorities was considered in [13], where the authors derived an exact method to estimate the steady state probabilities. In [2] the authors have studied the multi-class queueing system with abandonment. For the abandonment rate $\theta$ they described the $c\mu/\theta$-rule for in-queue cost minimization problem, i.e. when the customers in service do not incur the waiting cost. a similar problem with additional waiting costs was studied in [3]. The problem of identification and analysing the optimal scheduling policy that minimizes a range of cost functions of the system queue sizes, including the total number of queued packets, was the main objective in [1].

In contrast to most of previous papers, the customers in our system incurs a waiting cost in the queue which has significant contribution to a structure of an optimal allocation policy. The customers of a certain class can be served without any penalty at a server of the corresponding group. A cross-connectivity of customers to the servers of another group is also permitted but it can be performed according to some specified allocation policy and is resulting in additional usage cost. Our optimization and performance analysis of the queueing system includes the following contributions:

(a) We derive dynamic programming equations, develop policy-iteration algorithm and evaluate optimal allocation policy for given values of system parameters.
(b) We develop equations for the computation of any arbitrary moment of the busy period and the number of customers served for a fixed allocation policy.
(c) We obtain equations to calculate the distribution function for the maximal queue length during a busy period under a fixed allocation policy.

The rest of the paper is organized as follows. In Sect. 2, we describe the mathematical model. In Sect. 3, we formulate the Markov decision process and derive dynamic programming equations. The algorithmic analysis of the mean performance measures is given in Sect. 4. Section 5 is devoted to analysis of the maximal queue length in a busy period. Each section is supplemented with numerical examples.

We use the notations $\mathbf{e}_j(r)$ of dimension $r$ with 1 in the $j$th position and 0 elsewhere. The notation $1_{\{A\}}$ specifies the indicator function, which takes the value 1 if the event $A$ holds, and 0 otherwise.

## 2 Mathematical Model

We model the proposed system as a two-class queueing system with two heterogeneous groups of multiple servers (see Fig. 1). Let $N$ denote the number of servers. There are two classes of customers, which arrive to the system according to a Poisson process with mean $\lambda p_k$ $(0 < \lambda < \infty)$ for class $k \in \{1, 2\}$, $p_1 + p_2 = 1$. Each queue of class-$k$ customers has its own group of servers of size $N_k$, where $N_1 + N_2 = N$. The service times at server in a group $k$ are exponentially distributed with parameter $\mu_k$, with $0 < \mu_2 \leq \mu_1 < \infty$. The servers are assumed to be non preemptive, i.e. the service of a customer cannot be interrupted at any moment. We emphasise that all servers must be occupied whenever there are waiting customers. We include waiting costs $c_{k0} > 0$ for each unit time a customer of class $k$ spends in a queue. The customers of class $k$ can be served by a server of the own group resulting in the usage cost $c_{k1} > 0$ or by server of another group against extra charge $c_{k2} > c_{k1}$ per unit of time. The main goal is to allocate the servers between two queues with the aim to minimize the long-run average cost per unit of time.

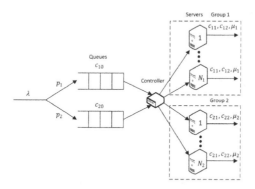

**Fig. 1.** Two-class queueing system with controllable cross-connectivity to different groups of servers

We formulate this task as a Markov decision problem and use dynamic programming approach to calculate performance measures under optimal control policy for correlated inter-arrival times. The main idea consists in calculating of optimal policy by constructing a sequence of improved policies that converges to the optimal one. It can be performed by solution of the system of Bellman's optimality equation [10, 11]. Let $Q_k(t) \in E_{Q_k} \equiv \mathbb{N}_0$ denote the number of customers in a queue $k$, $D_{kj}(t) \in E_{D_{kj}} = \{0, 1, \dots, N_k\}, k, j \in \{1, 2\}$, the number of busy servers of queue $k$ servicing the customers of queue $j$. The system states at time $t$ are described by a multi-dimensional continuous-time Markov chain

$$\{X(t)\}_{t \geq 0} = \{Q_1(t), Q_2(t), D_{11}(t), D_{12}(t), D_{21}(t), D_{22}(t)\}_{t \geq 0}. \tag{1}$$

with a state space $E_X$ defined as

$$E_X = \bigcup_{q_1=0}^{\infty} \bigcup_{q_2=0}^{\infty} E(q_1, q_2), \tag{2}$$

where

$$E(0, 0) = \{(0, 0, d_{11}, d_{12}, d_{21}, d_{22}) : d_{kj} \in E_{D_{kj}}, d_{k1} + d_{k2} \leq N_k\},$$
$$E(q_1, q_2) = \{(q_1, q_2, d_{11}, d_{12}, d_{21}, d_{22}) : q_1 + q_2 \geq 1, d_{kj} \in E_{D_{kj}}, d_{k1} + d_{k2} = N_k\}.$$

Further in the paper the notations $q_k(x)$ and $d_{kj}(x)$ will be used to specify the components of the vector state $x \in E_X$.

## 3   MBD Formulation

The controllable model associated with a Markov chain (1) is a five-tuple

$$\{E_X, A, \{A_{kj}(x), d_{kj}(x) > 0, q_1(x) + q_2(x) \geq 1\}, \lambda_{xy}(a), c(x, a)\}, \tag{3}$$

where

- $A = \{1, 2\}$ is an *action space* with elements $a \in A$, where $a = k$ means "to allocate a server to a queue $k$".
- $A_{kj}(x) \subseteq A$ for $d_{kj}(x) > 0, q_1(x) + q_2(x) \geq 1$, denotes *subset of actions* in state $x$ occurred at a decision epoch including the moments of time just after a service completion of a customer from queue $j$ at server $k$ in case when at least one of queue is non-empty, i.e.

$$A_{kj}(q_1, q_2, d_{11}, d_{12}, d_{21}, d_{22}) = \begin{cases} A, & q_1 \geq 1, q_2 \geq 1, \\ \{1\}, & q_1 \geq 1, q_2 = 0, \\ \{2\}, & q_1 = 0, q_2 \geq 1. \end{cases}$$

- $\lambda_{xy}(a)$ is a *transition rate* to go from $x$ to state $y$ under a control action $a$ associated with a state occurred just after an arrival in state $x$ or just after a service completion in state $x$. The model is conservative, i.e.

$$\lambda_{xy}(a) \geq 0, y \neq x, \lambda_{xx}(a) = -\lambda_x(a) = -\sum_{y \neq x} \lambda_{xy}(a), \lambda_x(a) < \infty,$$

where for $y \neq x$ and $\mathbf{e}_j := \mathbf{e}_j(6)$ we have

$$\lambda_{xy}(a) = \begin{cases} \lambda p_k, & y = x + \mathbf{e}_{2k+1}, \ \sum_{j=1}^2 d_{kj}(x) < N_k \\ \lambda p_k, & y = x + \mathbf{e}_{8-2k}, \ \sum_{j=1}^2 d_{kj}(x) = N_k, \\ & \sum_{j=1}^2 d_{(3-k)j}(x)(x) < N_{3-k}, \\ \lambda p_k, & y = x + \mathbf{e}_k, \ \sum_{j=1}^2 d_{kj}(x) = N_k, \\ \mu_k d_{kj}(x), & y = x - \mathbf{e}_{2k+j}, \ \sum_{k=1}^2 q_k(x) = 0, \\ \mu_k d_{kj}, & y = x - \mathbf{e}_a - \mathbf{e}_{2k+j} + \mathbf{e}_{(5-2a)k+(a-1)3}, \\ & \sum_{k=1}^2 q_k(x) \geq 1, \ a \in A_{kj}(x). \end{cases}$$

– $c(x, a)$ is an *immediate cost* in state $x \in E_X$ under control action $a \in A(x)$,

$$c(x, a) = c(x) = \sum_{k=1}^2 (c_{k0} q_k(x) + c_{k1} d_{k1}(x) + c_{k2} d_{k2}(x)).$$

The setting $c_{k0} = c_{k1} = c_{k2} = 1$ for $k \in \{1, 2\}$ leads to the number of customers in state $x$ which is independent of $a$.

A controller or decision maker chooses an action according to the following decision rule which will refer to as *stationary policy*.

**Definition 1.** *A stationary policy is a vector function* $f = (f_{11}, f_{12}, f_{21}, f_{22}) :$ $E_X \to (A_{11}(x), A_{12}(x), A_{21}(x), A_{22}(x))$ *which prescribes a selection of a control action* $a \in A_{kj}(x)$ *whenever the process* $\{X(t)\}_{t \geq 0}$ *is in state* $x \in E_X$ *just before a service completion of customer from queue* $j$ *at server* $k$ *if* $d_{kj} > 0$ *and* $q_1(x) + q_2(x) \geq 1$. *Any other transition cannot impute a selection of some action.*

For any fixed stationary policy $f$ we wish to guarantee that the process $\{X(t)\}_{t \geq 0}$ is an irreducible, positive recurrent Markov chain with a state space $E_X$ and infinitesimal generator $\Lambda_X$. As it is known, for ergodic Markov chains with costs the long-run average cost per unit of time for the policy $f$ exists [12] and coincides with corresponding assemble average,

$$g^f = \limsup_{t \to \infty} \frac{1}{t} V^f(x, t) = \sum_{y \in E_X} c(y) \pi_y^f, \tag{4}$$

where

$$V^f(x, t) = \mathbb{E}^f \Big[ \int_0^t \sum_{k=1}^2 (c_{k0} Q_k(t) + c_{k1} D_{k1}(t) + c_{k2} D_{k2}(t)) dt \, | X(0) = x \Big]$$

denotes the *total average cost* up to time $t$ given initial state is $x$ and $\pi_y^f = \mathbb{P}^f[X(t) = y]$ denotes a stationary state probability of the process under given policy $f$. The policy $f^*$ is said to be optimal when

$$g^* = \inf_f g^f. \tag{5}$$

One fruitful approach to finding optimal policy $f^*$ is through solving average cost optimality equation

$$Bv(x) = v(x) + g, \qquad (6)$$

where $B$ is a *dynamic programming operator* acting on value function $v : E_X \to \mathbb{R}$ which indicates a transient effect of an initial state $x$ to the total average cost and satisfies an asymptotic relation

$$V^f(x,t) = g^f t + v^f(x) + o(1), \; x \in E, \; t \to \infty. \qquad (7)$$

The optimality equation allows to construct a sequence of improved policies until the average cost optimal is reached. The functions $v^f$ and $g^f$ further in the paper will be denoted by $v$ and $g$ without upper index $f$.

**Theorem 1.** *The dynamic programming operator $B$ is defined as follows*

$$Bv(x) = \frac{1}{\lambda + \sum_{k=1}^{2} \sum_{j=1}^{2} \mu_k d_{kj}(x)} \Big[ c(x)$$

$$+ \lambda \sum_{k=1}^{2} p_k v(x + e_{2k+1}) 1_{\{\sum_{j=1}^{2} d_{kj}(x) < N_k\}},$$

$$+ \lambda \sum_{k=1}^{2} p_k v(x + e_{8-2k}) 1_{\{\sum_{j=1}^{2} d_{kj}(x) = N_k, \sum_{j=1}^{2} d_{(3-k)j}(x) < N_{3-k}\}}$$

$$+ \lambda \sum_{k=1}^{2} p_k v(x + e_k) 1_{\{\sum_{j=1}^{2} d_{kj}(x) = N_k\}}$$

$$+ \sum_{k=1}^{2} \sum_{j=1}^{2} \mu_k d_{kj}(x) v(x - e_{2k+j}) 1_{\{\sum_{k=1}^{2} q_k(x) = 0\}}$$

$$+ \sum_{k=1}^{2} \sum_{j=1}^{2} \mu_k d_{kj}(x) T_{kj} v(x) 1_{\{\sum_{k=1}^{2} q_k(x) \geq 1\}} \Big], \qquad (8)$$

*where operator $T_{kj}$ stands for decision making at decision epoch and is of the form*

$$T_{kj} v(x) = \min_{a \in A_{kj}(x)} v(x - e_a - e_{2k+j} + e_{(5-2a)k+(a-1)3}). \qquad (9)$$

*Proof.* According to [11] in general case we have

$$Bv(x) = \min_a \Big\{ \frac{1}{\lambda_x(a)} \Big[ c(x) + \sum_{y \neq x} \lambda_{xy}(a) v(y) \Big] \Big\}.$$

Evaluating these equations for analysed queueing system and taking into account the transition rates of the specified Markov decision model we get (8), where the term $c(x)$ is a number of customers in state $x \in E_X$, the next three terms represent the changing of the state accompanying with an arrival with a rate $\lambda$. The next two terms represent transitions due to service completions with rates $\mu_{kj}$ of customer from queue of class $j$ at server of class $k$. At a decision epoch a new state after transition is obtained obviously using a relation (9).

*Remark 1.* The infinite buffer queueing system is approximated by a finite buffer equivalent system. For bounded puffer size $q_k \leq B_k$ the size of the set space

$$|E_X| = \frac{(N_1 + 1)(N_1 + 2)(N_2 + 1)(N_2 + 2)}{4} + ((B_1 + 1)(B_2 + 1) - 1)(N_1 + 1)(N_2 + 1).$$

The buffer size $B_1 = B_2 = B$ will be chosen in such a way that it satisfies the condition

$$B > \frac{\log \varepsilon}{\log(\rho_1) + \log(\rho_2)}, \quad \rho_k = \frac{\lambda_k}{N_k \mu_k},$$

obtained from joint stationary state distribution of the number of customers in parallel independent queueing systems $M/M/N_1$ and $M/M/N_2$. The sum of probabilities $(1 - \rho_1)(1 - \rho_2) \sum_{i=B}^{\infty} \sum_{j=B}^{\infty} \rho_1^i \rho_2^j$ of states where the number of customers exceeds some level $B$ must be smaller as a predefined value $\varepsilon > 0$.

Further we convert the six dimensional state space $E_X$ of the Markov decision process ordered in a certain way to a one-dimensional equivalent state space $\mathbb{N}_0$, $\Delta : E_X \rightarrow \mathbb{N}_0$, for state $x = (q_1, q_2, d_{11}, d_{12}, d_{21}, d_{22}) \in E_X$

$$\Delta(x) = \left( d_{11}(x)[2(N_1 + 1) - d_{11}(x) + 1] + 2d_{12}(x) \right) \frac{(N_2 + 1)(N_2 + 2)}{2} \quad (10)$$

$$+ \left( 2(N_2 + 1) - d_{21} + 1 \right) \frac{d_{21}(x)}{2} + d_{22}(x), \quad q_1(x) = q_2(x) = 0,$$

$$\Delta(x) = \frac{(N_1 + 1)(N_1 + 2)(N_2 + 1)(N_2 + 2)}{4}$$

$$+ \left( q_1(x)(B_2 + 1) + q_2(x) - 1 \right) (N_1 + 1)(N_2 + 1)$$

$$+ d_{11}(x)(N_2 + 1) + d_{21}(x), \quad q_1(x) > 0 \| q_2(x) > 0.$$

In all algorithms below all appearing vector states must be transformed by a function (10), d.h. $x \mapsto \Delta(x), x \in E_X$.

*Example 1.* Consider the queueing system with arrival rates $\lambda = 20, 60, 100$. The number of servers in each group is $N_k = 5, k \in \{1, 2\}$. The system parameters take the following values

| $k$ | 1 | | 2 | | | |
|---|---|---|---|---|---|---|
| $p_k$ | 0.5 | | 0.5 | | | |
| $\mu_k$ | 20 | | 5 | | | |
| $c_{k0}, c_{k1}, c_{k2}$ | 20 | 0 | 25 | 5 | 0 | 10 |

By means of Algorithm 1 we calculate the optimized values of the long-run average cost, $g = 0.468, 19.984, 67.142$, respectively. The optimal control policy $f(x) = \{f_{kj}(x), k, j \in \{1, 2\}\}$ e.g. for states $x = (q_1, q_2, 1, 4, 4, 1)$ are summarized in Table 1. The regions where allocation of the queue 2 is optimal are marked by different grey colors dependent on arrival rate. The most dark region in tables corresponds to the lowest arrival rate. This experiment shows that the first group of servers has to server only the 1-class customers. The second group of servers has to serve the 1-class customers when there are only few customers in a second queue. Upon increasing of the second queue length or/and of the arrival rate the incentive to make an allocation of a queue 2 to servers from the second group is getting higher.

---

**Algorithm 1.** Policy iteration algorithm

---

1: **procedure** MDP$(\lambda, p_k, \mu_{kj}, c_{0k}, c_{kj}, k, j \in \{1, 2\})$

2:    $f_{kj}^{(0)}(x) = \begin{cases} k, & q_k(x) > 0, d_{kj}(x) > 0, \\ 3 - k, & q_k(x) = 0, q_{3-k}(x) > 0, d_{kj}(x) > 0. \end{cases}$          ▷ Initial policy

3:    $n \leftarrow 0$

4:    $g^{(n)} = \lambda \sum_{k=1}^{2} p_k v^{(n)}(e_{2k+1})$          ▷ Evaluation of value function

5:    **for** $x = (0, 0, 0, 0, 0, 1)$ **to** $(B_1, B_2, N_1, 0, N_2, 0)$ **do**

6:

$$v^{(n)}(x) = \frac{1}{\lambda + \sum_{k=1}^{2} \sum_{j=1}^{2} \mu_k d_{kj}(x)} \Big[ c(x) - g^{(n)} + \lambda \sum_{k=1}^{2} p_k v^{(n)}(x + e_{2k+1}) 1_{\{\sum_{j=1}^{2} d_{kj}(x) < N_k\}}$$

$$+ \lambda \sum_{k=1}^{2} p_k v^{(n)}(x + e_{8-2k}) 1_{\{\sum_{j=1}^{2} d_{kj}(x) = N_k, \sum_{j=1}^{2} d_{(3-k)j}(x) < N_{3-k}\}} + \lambda \sum_{k=1}^{2} p_k v^{(n)}(x) 1_{\{q_k(x) = B_k\}}$$

$$+ \lambda \sum_{k=1}^{2} p_k v^{(n)}(x + e_k) 1_{\{\sum_{k=1}^{2} \sum_{j=1}^{2} d_{kj}(x) = N\}} + \sum_{k=1}^{2} \sum_{j=1}^{2} \mu_k d_{kj}(x) v^{(n)}(x - e_{2k+j}) 1_{\{\sum_{k=1}^{2} q_k(x) = 0\}}$$

$$+ \sum_{k=1}^{2} \sum_{j=1}^{2} \mu_k d_{kj}(x) v^{(n)}(x - e_{f_{kj}^{(n)}(x)} - e_{2k+j} + e_{(5-2f_{kj}^{(n)}(x))k + (f_{kj}^{(n)}(x)-1)3}) 1_{\{\sum_{k=1}^{2} q_k(x) \geq 1\}} \Big]$$

7:    **end for**

8:          ▷ Policy improvement

$$f_{kj}^{(n+1)}(x) = \operatorname{argmin}_{a \in A_{kj}(x)} v^{(n)}(x - e_a - e_{2k+j} + e_{(5-2a)k + (a-1)3})$$

9:    **if** $f_{kj}^{(n+1)}(x) = f_{kj}^{(n)}$, $x \in E_X$ **then return** $f_{kj}^{(n+1)}(x), v^{(n)}(x), g^{(n)}$

10:    **else** $n \leftarrow n + 1$, go to step 4

11:    **end if**

12: **end procedure**

---

*Example 2.* In this example by means of a Table 2 we list the optimal number of busy servers in each group servicing the first class customers dependent on queue lengths under the assumption of possible service preemption. We observe that while for short queues the cross-allocation makes sense, for large queue lengths it is always better to allocate all servers in a certain group to the own class of customers and not to use a cross-connectivity. Of course the corresponding thresholds are strongly dependent on the values of waiting and usage costs.

**Table 1.** Optimal control policies $f_{kj}(q_1, q_2, 1, 4, 4, 1)$

| $f_{11}$ | 0 | 1 | 2 | 3 | 4 | 5 | 6 | 7 | 8 | 9 | ... $q_2$ |
|---|---|---|---|---|---|---|---|---|---|---|---|
| 0 | - | 2 | 2 | 2 | 2 | 2 | 2 | 2 | 2 | 2 | ... |
| 1 | 1 | 1 | 1 | 1 | 1 | 1 | 1 | 1 | 1 | 1 | ... |
| 2 | 1 | 1 | 1 | 1 | 1 | 1 | 1 | 1 | 1 | 1 | ... |
| 3 | 1 | 1 | 1 | 1 | 1 | 1 | 1 | 1 | 1 | 1 | ... |
| 4 | 1 | 1 | 1 | 1 | 1 | 1 | 1 | 1 | 1 | 1 | ... |
| 5 | 1 | 1 | 1 | 1 | 1 | 1 | 1 | 1 | 1 | 1 | ... |
| 6 | 1 | 1 | 1 | 1 | 1 | 1 | 1 | 1 | 1 | 1 | ... |
| 7 | 1 | 1 | 1 | 1 | 1 | 1 | 1 | 1 | 1 | 1 | ... |
| 8 | 1 | 1 | 1 | 1 | 1 | 1 | 1 | 1 | 1 | 1 | ... |
| 9 | 1 | 1 | 1 | 1 | 1 | 1 | 1 | 1 | 1 | 1 | ... |
| $q_1$ | | | | | | | | | | | |

| $f_{12}$ | 0 | 1 | 2 | 3 | 4 | 5 | 6 | 7 | 8 | 9 | ... $q_2$ |
|---|---|---|---|---|---|---|---|---|---|---|---|
| 0 | - | 2 | 2 | 2 | 2 | 2 | 2 | 2 | 2 | 2 | ... |
| 1 | 1 | 1 | 1 | 1 | 1 | 1 | 1 | 1 | 1 | 1 | ... |
| 2 | 1 | 1 | 1 | 1 | 1 | 1 | 1 | 1 | 1 | 1 | ... |
| 3 | 1 | 1 | 1 | 1 | 1 | 1 | 1 | 1 | 1 | 1 | ... |
| 4 | 1 | 1 | 1 | 1 | 1 | 1 | 1 | 1 | 1 | 1 | ... |
| 5 | 1 | 1 | 1 | 1 | 1 | 1 | 1 | 1 | 1 | 1 | ... |
| 6 | 1 | 1 | 1 | 1 | 1 | 1 | 1 | 1 | 1 | 1 | ... |
| 7 | 1 | 1 | 1 | 1 | 1 | 1 | 1 | 1 | 1 | 1 | ... |
| 8 | 1 | 1 | 1 | 1 | 1 | 1 | 1 | 1 | 1 | 1 | ... |
| 9 | 1 | 1 | 1 | 1 | 1 | 1 | 1 | 1 | 1 | 1 | ... |
| $q_1$ | | | | | | | | | | | |

| $f_{21}$ | 0 | 1 | 2 | 3 | 4 | 5 | 6 | 7 | 8 | 9 | ... $q_2$ |
|---|---|---|---|---|---|---|---|---|---|---|---|
| 0 | - | 2 | 2 | 2 | 2 | 2 | 2 | 2 | 2 | 2 | ... |
| 1 | 1 | 1 | 2 | 2 | 2 | 2 | 2 | 2 | 2 | 2 | ... |
| 2 | 1 | 1 | 2 | 2 | 2 | 2 | 2 | 2 | 2 | 2 | ... |
| 3 | 1 | 1 | 1 | 2 | 2 | 2 | 2 | 2 | 2 | 2 | ... |
| 4 | 1 | 1 | 1 | 2 | 2 | 2 | 2 | 2 | 2 | 2 | ... |
| 5 | 1 | 1 | 1 | 2 | 2 | 2 | 2 | 2 | 2 | 2 | ... |
| 6 | 1 | 1 | 1 | 2 | 2 | 2 | 2 | 2 | 2 | 2 | ... |
| 7 | 1 | 1 | 1 | 2 | 2 | 2 | 2 | 2 | 2 | 2 | ... |
| 8 | 1 | 1 | 1 | 1 | 2 | 2 | 2 | 2 | 2 | 2 | ... |
| 9 | 1 | 1 | 1 | 1 | 1 | 2 | 2 | 2 | 2 | 2 | ... |
| $q_1$ | | | | | | | | | | | |

| $f_{22}$ | 0 | 1 | 2 | 3 | 4 | 5 | 6 | 7 | 8 | 9 | ... $q_2$ |
|---|---|---|---|---|---|---|---|---|---|---|---|
| 0 | - | 2 | 2 | 2 | 2 | 2 | 2 | 2 | 2 | 2 | ... |
| 1 | 1 | 2 | 2 | 2 | 2 | 2 | 2 | 2 | 2 | 2 | ... |
| 2 | 1 | 2 | 2 | 2 | 2 | 2 | 2 | 2 | 2 | 2 | ... |
| 3 | 1 | 1 | 2 | 2 | 2 | 2 | 2 | 2 | 2 | 2 | ... |
| 4 | 1 | 1 | 2 | 2 | 2 | 2 | 2 | 2 | 2 | 2 | ... |
| 5 | 1 | 1 | 1 | 2 | 2 | 2 | 2 | 2 | 2 | 2 | ... |
| 6 | 1 | 1 | 1 | 2 | 2 | 2 | 2 | 2 | 2 | 2 | ... |
| 7 | 1 | 1 | 1 | 2 | 2 | 2 | 2 | 2 | 2 | 2 | ... |
| 8 | 1 | 1 | 1 | 2 | 2 | 2 | 2 | 2 | 2 | 2 | ... |
| 9 | 1 | 1 | 1 | 1 | 2 | 2 | 2 | 2 | 2 | 2 | ... |
| $q_1$ | | | | | | | | | | | |

# 4   Mean Performance Measures

The Policy iteration Algorithm 1 can be easily adopted to evaluation of different mean performance measures. Due to the fact that for a fixed control policy $f$ $g^f = \sum_{x \in E_X} c(x)\pi_x^f$, the evaluation step of the value function can be repeated with a corresponding changing of the immediate cost function $c(x)$.

**Corollary 1.** *Mean number of class-m customers in a queue:*

$$\bar{Q}_m = \sum_{x \in E_X} c(x)\pi_x^f = \sum_{q_1=0}^{\infty} \sum_{q_2=0}^{\infty} \sum_{d_{11}+d_{12}=N_1} \sum_{d_{21}+d_{22}=N_2} q_m \pi_{(q_1,q_2,d_{11},d_{12},d_{21},d_{22})}^f,$$

*where $c(x) = q_m(x), x \in E_X$.*
*Mean number of busy servers with class-m customers:*

$$\bar{C}_m = \sum_{x \in E_X} c(x)\pi_x^f = \sum_{q_1=0}^{\infty} \sum_{q_2=0}^{\infty} \sum_{k=1}^{2} \sum_{j=1}^{2} (d_{m1} + d_{m2})\pi_{(q_1,q_2,d_{11},d_{12},d_{21},d_{22})}^f,$$

*where $c(x) = \sum_{j=1}^{2} d_{mj}(x)$.*

**Table 2.** Optimal number of busy servers in group 1 and group 2 with 1-class customers

| $d_{11}$ | 1 | 2 | 3 | 4 | 5 | 6 | 7 | 8 | 9 | $\dots q_2$ | $d_{22}$ | 1 | 2 | 3 | 4 | 5 | 6 | 7 | 8 | 9 | $\dots q_2$ |
|---|---|---|---|---|---|---|---|---|---|---|---|---|---|---|---|---|---|---|---|---|---|
| 1 | 2 | 4 | 5 | 5 | 5 | 5 | 5 | 5 | 5 | ... | 1 | 3 | 3 | 3 | 3 | 0 | 0 | 0 | 0 | 0 | ... |
| 2 | 2 | 5 | 5 | 5 | 5 | 5 | 5 | 5 | 5 | ... | 2 | 3 | 3 | 3 | 3 | 0 | 0 | 0 | 0 | 0 | ... |
| 3 | 3 | 5 | 5 | 5 | 5 | 5 | 5 | 5 | 5 | ... | 3 | 3 | 3 | 3 | 0 | 0 | 0 | 0 | 0 | 0 | ... |
| 4 | 3 | 5 | 5 | 5 | 5 | 5 | 5 | 5 | 5 | ... | 4 | 2 | 3 | 3 | 0 | 0 | 0 | 0 | 0 | 0 | ... |
| 5 | 5 | 5 | 5 | 5 | 5 | 5 | 5 | 5 | 5 | ... | 5 | 2 | 2 | 0 | 0 | 0 | 0 | 0 | 0 | 0 | ... |
| 6 | 5 | 5 | 5 | 5 | 5 | 5 | 5 | 5 | 5 | ... | 6 | 2 | 2 | 0 | 0 | 0 | 0 | 0 | 0 | 0 | ... |
| 7 | 5 | 5 | 5 | 5 | 5 | 5 | 5 | 5 | 5 | ... | 7 | 2 | 0 | 0 | 0 | 0 | 0 | 0 | 0 | 0 | ... |
| 8 | 5 | 5 | 5 | 5 | 5 | 5 | 5 | 5 | 5 | ... | 8 | 0 | 0 | 0 | 0 | 0 | 0 | 0 | 0 | 0 | ... |
| 9 | 5 | 5 | 5 | 5 | 5 | 5 | 5 | 5 | 5 | ... | 9 | 0 | 0 | 0 | 0 | 0 | 0 | 0 | 0 | 0 | ... |
| ⋮ | ⋮ | ⋮ | ⋮ | ⋮ | ⋮ | ⋮ | ⋮ | ⋮ | ⋮ | ⋱ | ⋮ | ⋮ | ⋮ | ⋮ | ⋮ | ⋮ | ⋮ | ⋮ | ⋮ | ⋮ | ⋱ |
| $q_1$ | | | | | | | | | | | $q_1$ | | | | | | | | | | |

*Mean number of customers in the system*

$$\bar{N} = \sum_{x \in E_X} c(x)\pi_x^f = \sum_{m=1}^{2}(\bar{Q}_m + \bar{C}_m),$$

where $c(x) = \sum_{m=1}^{2}[q_m(x) + \sum_{j=1}^{2} d_{mj}(x)]$.

*Example 3.* Here we fix $\mu_1 = 20, \mu_2 = 5$ and vary $\lambda$. The system operates under optimal policy for the average cost criterion. Table 3 lists values of $\bar{Q}_m, \bar{C}_m$, and $\bar{N}$ calculated by a modified Policy iteration Algorithm 1.

**Table 3.** $\bar{Q}_m, \bar{C}_m, m \in \{1, 2\}$, and $\bar{N}$ as $\lambda$ varies

| | $\lambda = 20$ | $\lambda = 60$ | $\lambda = 100$ | $\lambda = 120$ |
|---|---|---|---|---|
| $\bar{Q}_1$ | 0.000 | 0.023 | 0.422 | 1.143 |
| $\bar{Q}_2$ | 0.000 | 0.104 | 2.128 | 12.549 |
| $\bar{C}_1$ | 0.518 | 2.009 | 3.726 | 4.622 |
| $\bar{C}_2$ | 1.927 | 3.874 | 4.575 | 4.847 |
| $\bar{N}$ | 2.445 | 6.009 | 10.854 | 23.268 |

The busy period $L$ of the queueing system is the duration starting when an arriving customer of any class finds the system empty, i.e. the arriving customer enters the system at state $0 = \Delta(0, 0, 0, 0, 0, 0)$ and finishes when the system visits this empty state again at a service completion at any server. The following notation is introduced. Denote by $\tilde{\varphi}_x(s) = \int_0^\infty \varphi_x(t)e^{-st}dt, \mathrm{Re}[s] > 0$, the Laplace-Stieltjes transform (LST) of the first passage time to state 0 give that the initial state is $x \in E_x$ with a PDF $\varphi_x(t)$ and by $\bar{\Phi}(x)$ the corresponding first moment. According to the first step analysis the LST $\tilde{\varphi}_x(s)$ satisfy

$$\tilde{\varphi}_0(s) = 1,$$

$$\tilde{\varphi}_x(s) = \sum_{y \neq x} \frac{\lambda_{xy}(a)}{s + \lambda_x(a)} \tilde{\varphi}_y(s), \; x \in E_X \setminus \{0\}, \tag{11}$$

where $\lambda_x(a) = \sum_{y \neq x} \lambda_{xy}(a)$. Taking into account the property $\bar{\Phi}(x) = -\frac{d}{ds}\tilde{\varphi}_x(s)\Big|_{s=0}$, (11) can be modified to the system

$$\bar{\Phi}(0) = 1, \tag{12}$$

$$\bar{\Phi}(x) = \frac{1}{\lambda_x(a)}\Big[1 + \sum_{y \neq x} \lambda_{xy}(a)\bar{\Phi}(y)\Big], \; x \in E_X \setminus \{0\}.$$

The busy period in our system starts by visiting of states $(0,0,1,0,0,0)$ and $(0,0,0,0,1,0)$ with probabilities $p_1$ and $p_2$, hence the unconditional first moment

$$\bar{L} = \mathbb{E}[L] = p_1\bar{\Phi}(0,0,1,0,0,0) + p_2\bar{\Phi}(0,0,0,0,1,0). \tag{13}$$

*Remark 2.* Let $\bar{\Phi}^{(n)}$ the $n$th moment of the first passage time to state 0 give that the initial state is $x$. With the help of Leibnitz's formula for the derivative of a product, we differentiate the expression given in (11) to get

$$(s + \lambda_x(a))\frac{d^{(n)}\tilde{\varphi}_x(s)}{ds^n} + n\frac{d^{n-1}\tilde{\varphi}_x(s)}{ds^{n-1}} = \sum_{y \neq x} \lambda_{xy}(a)\frac{d^{(n)}\tilde{\varphi}_y(s)}{ds^n}.$$

---

**Algorithm 2.** Mean busy period

---

1: **procedure** MBP($\lambda, p_k, \mu_{kj}, f_{ij}(x)\, k, j \in \{1,2\}$)
2:     **for** $x = (0,0,0,0,0,1)$ **to** $(B_1, B_2, N_1, 0, N_2, 0)$ **do**
3:

$$\bar{\Phi}(x) = \frac{1}{\lambda + \sum_{k=1}^{2}\sum_{j=1}^{2}\mu_k d_{kj}(x)}\Big[1 + \lambda\sum_{k=1}^{2}p_k\bar{\Phi}(x + e_{2k+1})1_{\{\sum_{j=1}^{2}d_{kj}(x)<N_k\}}$$

$$+ \lambda\sum_{k=1}^{2}p_k\bar{\Phi}(x + e_{8-2k})1_{\{\sum_{j=1}^{2}d_{kj}(x)=N_k,\sum_{j=1}^{2}d_{(3-k)j}(x)<N_{3-k}\}} + \lambda\sum_{k=1}^{2}p_k\bar{\Phi}(x)1_{\{q_k(x)=B_k\}}$$

$$+ \lambda\sum_{k=1}^{2}p_k\bar{\Phi}(x + e_k)1_{\{\sum_{k=1}^{2}\sum_{j=1}^{2}d_{kj}(x)=N\}} + \sum_{k=1}^{2}\sum_{j=1}^{2}\mu_k d_{kj}(x)\bar{\Phi}(x - e_{2k+j})1_{\{\sum_{k=1}^{2}q_k(x)=0\}}$$

$$+ \sum_{k=1}^{2}\sum_{j=1}^{2}\mu_k d_{kj}(x)\bar{\Phi}(x - e_{f_{kj}(x)} - e_{2k+j} + e_{(5-2f_{kj}(x))k+(f_{kj}(x)-1)3})1_{\{\sum_{k=1}^{2}q_k(x)\geq1\}}\Big]$$

4:     **end for**
5:     **return** $\bar{L} = p_1\bar{\Phi}(0,0,1,0,0,0) + p_2\bar{\Phi}(0,0,0,0,1,0)$
6: **end procedure**

---

Since $\bar{\Phi}^{(n)}(x) = (-1)^n \frac{d}{ds}\tilde{\varphi}_x(s)\Big|_{s=0}$ we notice that

$$\bar{L}^{(n)} = p_1\bar{\Phi}^{(n)}(0,0,1,0,0,0) + p_2\bar{\Phi}^{(n)}(0,0,0,0,1,0),$$

where $\bar{\Phi}^{(n)}(x)$ can be calculated recursively by

$$\lambda_x(a)\bar{\Phi}^{(n)}(x) - \sum_{y\neq x}\lambda_{xy}(a)\bar{\Phi}^{(n)}(y) = n\bar{\Phi}^{(n-1)}(x).$$

---

**Algorithm 3.** Mean number of class-$m$ customers served during a busy period

1: **procedure** MNSC($\lambda, p_k, \mu_k, f_{ij}(x)\,k, j \in \{1,2\}$)
2:    $\bar{\psi}(0) = 1$
3:    **for** $x = (0,0,0,0,0,1)$ to $(B_1, B_2, N_1, 0, N_2, 0)$ **do**
4:

$$\bar{\Psi}_m(x) = \frac{1}{\lambda + \sum_{k=1}^{2}\sum_{j=1}^{2}\mu_k d_{kj}(x)}\Big[\mu_m\sum_{j=1}^{2}d_{mj}(x) + \lambda\sum_{k=1}^{2}p_k\bar{\Psi}_m(x + e_{2k+1})1_{\{\sum_{j=1}^{2}d_{kj}(x)<N_k\}}$$

$$+ \lambda\sum_{k=1}^{2}p_k\bar{\Psi}_m(x + e_{8-2k})1_{\{\sum_{j=1}^{2}d_{kj}(x)=N_k,\sum_{j=1}^{2}d_{(3-k)j}(x)<N_{3-k}\}} + \lambda\sum_{k=1}^{2}p_k\bar{\Psi}_m(x)1_{\{q_k(x)=B_k\}}$$

$$+ \lambda\sum_{k=1}^{2}p_k\bar{\Psi}_m(x + e_k)1_{\{\sum_{k=1}^{2}\sum_{j=1}^{2}d_{kj}(x)=N\}} + \sum_{k=1}^{2}\sum_{j=1}^{2}\mu_k d_{kj}(x)\bar{\Psi}_m(x - e_{2k+j})1_{\{\sum_{k=1}^{2}q_k(x)=0\}}$$

$$+ \sum_{k=1}^{2}\sum_{j=1}^{2}\mu_k d_{kj}(x)\bar{\Psi}_m(x - e_{f_{kj}(x)} - e_{2k+j} + e_{(5-2f_{kj}(x))k+(f_{kj}(x)-1)3})1_{\{\sum_{k=1}^{2}q_k(x)\geq 1\}}\Big]$$

5:    **end for**
6:    **return** $\bar{l}_m = p_1\bar{\Psi}_m(0,0,1,0,0,0) + p_2\bar{\Psi}_m(0,0,0,0,1,0)$
7: **end procedure**

---

*Example 4.* Algorithm 2 was implemented to calculate the mean busy period. We fix $\mu_2 = 5$, $p_k = 0.5, k \in \{1,2\}$. In Table 4 we display the mean busy period $\bar{L}$ for several choices of $\mu_1$ and $\rho$ for the system under optimal control policy for the average cost criterion. The measure $\bar{L}$ for fixed $\mu_1$ increases with increasing values of $\rho$. For fixed large $\rho$ and increasing $\mu$ the busy period is monotonously increasing, since in this case the higher values of arrival rate influence considerable the contribution of the second queue to the total busy period of the system. In case of small $\rho$ with increasing values of $\mu_1$ the values of $\bar{L}$ are not increasing.

**Table 4.** $\bar{L}$ as $\mu_1$ and $\rho$ vary

| $\rho$ | $\mu_1 = 5$ | $\mu_1 = 10$ | $\mu_1 = 15$ | $\mu_1 = 20$ |
|---|---|---|---|---|
| 0.1 | 0.344 | 0.277 | 0.279 | 0.301 |
| 0.3 | 1.273 | 1.230 | 1.651 | 2.282 |
| 0.6 | 13.794 | 15.701 | 26.042 | 44.019 |
| 0.8 | 88.332 | 101.771 | 170.474 | 285.996 |

Now we briefly discuss the evaluation of the mean number of customers served during a busy period in the queueing system under optimal control policy. Denote by $I_m$ the number of class-$m$ customers served during a busy period $L$. Further, denote by $\tilde{\psi}_{m,x}(z) = \sum_{i=1}^{\infty} \psi_{m,x}(i)z^i, |z| \le 1$, the probability generating function (PGF) of the PDF $\psi_{m,x}(i)$ of the number of service completions of class-$m$ customers up to the end of busy period given that the initial state is $x \in E_X \backslash \{0\}$. According to the law of the total probability, the density $\psi_{m,x}(i)$ satisfies the following system:

---

**Algorithm 4.** Maximal queue length in a busy period

1: **procedure** MQL($\lambda, p_k, \mu_{kj}, n_k, f_{ij}(x)\, k, j \in \{1,2\}$)
2:    $\tau(0) = 1, \tau(x) = 0, x \in E_{max}$
3:    **for** $x = (0,0,0,0,0,1)$ **to** $(n_1, n_2, N_1, 0, N_2, 0)$ **do**
4:

$$\tau(x) = \frac{1}{\lambda + \sum_{k=1}^{2}\sum_{j=1}^{2}\mu_k d_{kj}(x)}\Big[\lambda \sum_{k=1}^{2} p_k \tau(x + e_{2k+1})1_{\{\sum_{j=1}^{2} d_{kj}(x) < N_k\}}$$

$$+ \lambda \sum_{k=1}^{2} p_k \tau(x + e_{8-2k})1_{\{\sum_{j=1}^{2} d_{kj}(x) = N_k, \sum_{j=1}^{2} d_{(3-k)j}(x) < N_{3-k}\}}$$

$$+ \lambda \sum_{k=1}^{2} p_k \tau(x + e_k)1_{\{\sum_{k=1}^{2}\sum_{j=1}^{2} d_{kj}(x) = N\}} + \sum_{k=1}^{2}\sum_{j=1}^{2}\mu_k d_{kj}(x)\tau(x - e_{2k+j})1_{\{\sum_{k=1}^{2} q_k(x) = 0\}}$$

$$+ \sum_{k=1}^{2}\sum_{j=1}^{2}\mu_k d_{kj}(x)\tau(x - e_{f_{kj}(x)} - e_{2k+j} + e_{(5-2f_{kj}(x))k+(f_{kj}(x)-1)3})1_{\{\sum_{k=1}^{2} q_k(x) \ge 1\}}\Big]$$

5:    **end for**
6:    **return** $p_{max} = p_1 \tau(0,0,1,0,0,0) + p_2 \tau(0,0,0,0,1,0)$
7: **end procedure**

---

$$\psi_{m,x}(i) = \frac{\lambda_{xy'}(a)}{\lambda_x(a)}\psi_{m,y'}(i-1) + \sum_{y \ne x, y'}\frac{\lambda_{xy}(a)}{\lambda_x(a)}\psi_{m,y}(i). \qquad (14)$$

The first term on the right-hand side of (14) stands for the transition to the state $y'$, which we count (i.e., service completion of a class-$m$ customer), while the second term includes other possible transitions. In terms of the PGF, the last system can be rewritten as follows:

$$\tilde{\psi}_{m,x}(z) = \frac{z\lambda_{xy'}(a)}{\lambda_x(a)}\tilde{\psi}_{m,y'}(z) + \sum_{y \ne x, y'}\frac{\lambda_{xy}(a)}{\lambda_x(a)}\tilde{\psi}_{m,y}(z). \qquad (15)$$

The expressions for evaluating PGFs $\tilde{\psi}_{m,x}(z)$ proposed above can be modified to calculate the first moments $\bar{\Psi}_m(x) = \left.\frac{d\tilde{\psi}_{m,x}(z)}{dz}\right|_{z=1}$,

$$\bar{\Psi}_m(0) = 1,$$

$$\bar{\Psi}_m(x) = \frac{1}{\lambda_x}\Big[\lambda_{xy'}(a) + \sum_{y \ne x}\lambda_{xy}(a)\bar{\Psi}_m(y)\Big], \quad x \in E_X \backslash \{0\}, \ m \in \{1,2\}. \qquad (16)$$

For unconditional moment $\bar{I}_m$ we get

$$\bar{I}_m = \mathbb{E}[I_m] = p_1 \bar{\Psi}_m(0,0,1,0,0,0) + p_2 \bar{\Psi}_m(0,0,0,0,1,0). \tag{17}$$

For the system under study these expressions are given in Algorithm 3.

*Remark 3.* By differentiating the expression in (15), we find that

$$\frac{d^n \tilde{\psi}_{m,x}(z)}{dz^n} = \frac{\lambda_{xy'}(a)}{\lambda_x(a)} \left[ z \frac{d^n \tilde{\psi}_{m,y'}(z)}{dz^n} + n \frac{d^{n-1} \tilde{\psi}_{m,y'}(z)}{dz^{n-1}} \right] + \sum_{y \neq x, y'} \frac{\lambda_{xy}(a)}{\lambda_x(a)} \frac{d^n \tilde{\psi}_{m,y}(z)}{dz^n}.$$

Noting that $\bar{\Psi}_m^{(n)}(x) = \left. \frac{d^n \tilde{\psi}_{m,x}(z)}{dz^n} \right|_{z=1}$ we get an appropriate formula for computing any arbitrary unconditional factorial moment

$$\bar{I}_m^{(n)} = \mathbb{E}[I_m(I_m - 1)\dots(I_m - n + 1)] = p_1 \bar{\Psi}_m^{(n)}(0,0,1,0,0,0) + p_2 \bar{\Psi}_m^{(n)}(0,0,0,0,1,0),$$

where

$$\bar{\Psi}_m^{(n)}(x) - \sum_{y \neq x} \frac{\lambda_{xy}(a)}{\lambda_x(a)} \bar{\Psi}_m^{(n)}(y) = n \frac{\lambda_{xy'}(a)}{\lambda_x(a)} \bar{\Psi}_m^{(n-1)}(y').$$

*Example 5.* Table 5 lists values of the mean of $I_m, m \in \{1,2\}$, versus $\rho$ and $\mu_1$. From this table and Table 4, it can be verified the identity $\sum_{m=1}^{2} \bar{I} = 1 + \lambda \bar{L}$. Here both measures increase with increasing values of $\rho$ and $\mu_1$.

**Table 5.** $\bar{I}_1$ and $\bar{I}_2$ as $\mu_1$ and $\rho$ vary

| $\rho$ | $\mu_1 = 5$ | $\mu_1 = 10$ | $\mu_1 = 15$ | $\mu_1 = 20$ | $\rho$ | $\mu_1 = 5$ | $\mu_1 = 10$ | $\mu_1 = 15$ | $\mu_1 = 20$ |
|---|---|---|---|---|---|---|---|---|---|
| 0.1 | 1.359 | 1.540 | 1.897 | 2.381 | 0.1 | 1.359 | 1.539 | 1.892 | 2.372 |
| 0.3 | 10.053 | 14.363 | 25.352 | 45.410 | 0.3 | 10.000 | 14.202 | 24.863 | 44.197 |
| 0.6 | 211.496 | 367.328 | 822.890 | 1752.920 | 0.6 | 204.981 | 331.841 | 709.828 | 1466.680 |
| 0.8 | 1827.370 | 3235.500 | 7303.150 | 15360.300 | 0.8 | 1725.070 | 2768.290 | 5954.750 | 12254.000 |

## 5    Maximal Queue Length in a Busy Period

We now provide the algorithmic analysis of the maximum length of queues observed during a busy period $L$ of the queueing system under optimal control policy which minimizes the long-run average cost $g$. Denote by $B_{k,max}, k \in \{1,2\}$, the maximum number of class-$k$ customers waiting in the queue during a busy period. For each fixed value $n_k \geq 1$ the event $\{B_{k,max} < n_k, k \in \{1,2\}\}$ is equivalent to the event that, starting from the states $(0,0,1,0,0,0)$ with probability $p_1$ and $(0,0,0,0,1,0)$ with probability $p_2$, the process $\{X(t)\}_{t \geq 0}$ hits the empty state $0 = (0,0,0,0,0,0)$ before visiting the subset of states

$$E_{max} = \cup_{q_1=0}^{n_1} E(q_1, n_2 + 1) \cup_{q_2=0}^{n_2} E(n_1 + 1, q_2).$$

We compute $p_{max} = \mathbb{P}[B_{1,max} \leq n_1, B_{2,max} \leq n_2]$ by means of probabilities of absorption in one of state in

$$E_{max} \cup \{(0,0,0,0,0,0)\}$$

given that the initial state is $x \in \cup_{q_1=0}^{n_1} \cup_{q_2=0}^{n_2} E(q_1, q_2)$. Denote by $\tau(x)$ the probability of absorption into empty state $0$ starting from $x$. By conditioning on the next visited state, we get

$$\tau(0) = 1,$$

$$\tau(x) = \frac{1}{\lambda_x(a)} \sum_{y \neq x} \lambda_{xy}(a)\tau(y),$$

$$\tau(x) = 0, \; x \in E_{max}.$$

**Table 6.** $\mathbb{P}[B_{1,max} \leq n_1]$ and $\mathbb{P}[B_{2,max} \leq n_2]$ as $\lambda$ varies

| $n_1$ | $\lambda = 20$ | $\lambda = 60$ | $\lambda = 80$ | $\lambda = 100$ | $n_2$ | $\lambda = 20$ | $\lambda = 60$ | $\lambda = 80$ | $\lambda = 100$ |
|---|---|---|---|---|---|---|---|---|---|
| 0 | 0.99972 | 0.30435 | 0.18305 | 0.13879 | 0 | 0.99968 | 0.29782 | 0.18239 | 0.13869 |
| 1 | 0.99997 | 0.38986 | 0.18944 | 0.13948 | 1 | 0.99995 | 0.35033 | 0.18620 | 0.139117 |
| 2 | 0.99999 | 0.57884 | 0.20533 | 0.14074 | 2 | 0.99999 | 0.46221 | 0.19358 | 0.13981 |
| 3 | 0.99999 | 0.80643 | 0.24466 | 0.14310 | 3 | 0.99999 | 0.54699 | 0.20064 | 0.14042 |
| 4 | 1 | 0.93660 | 0.33281 | 0.14760 | 4 | 0.99999 | 0.54290 | 0.20388 | 0.14075 |
| 5 | 1 | 0.982065 | 0.49060 | 0.15629 | 5 | 0.99999 | 0.73425 | 0.22708 | 0.14214 |
| ⋮ | ⋮ | ⋮ | ⋮ | ⋮ | ⋮ | ⋮ | ⋮ | ⋮ | ⋮ |
| 10 | 1 | 0.99996 | 0.98951 | 0.48731 | 10 | 1 | 0.98349 | 0.46128 | 0.15139 |
| ⋮ | ⋮ | ⋮ | ⋮ | ⋮ | ⋮ | ⋮ | ⋮ | ⋮ | ⋮ |
| 20 | 1 | 0.99998 | 0.99987 | 0.94812 | 20 | 1 | 0.99997 | 0.98851 | 0.26974 |
| ⋮ | ⋮ | ⋮ | ⋮ | ⋮ | ⋮ | ⋮ | ⋮ | ⋮ | ⋮ |
| 40 | 1 | 1 | 0.99999 | 0.94821 | 40 | 1 | 1 | 0.99999 | 0.94821 |

*Example 6.* Consider the queueing system with system parameters from Example 1 functioning under the optimal control policy for the average cost criterion. In this computational experiment we use Algorithm 4 and it's small modification for marginal distribution functions. In Table 6 and 7 the arrival rate is varied while other parameters were kept constant. Entries in Table 6 and 7 summarize a distribution function $\mathbb{P}[B_{k,max} \leq n_k]$ of maximal $k$th queue length as well as a joint distribution function $\mathbb{P}[B_{1,max} \leq n_1, B_{2,max} \leq n_2]$ for both of queues, respectively. As it can be expected we observe that these distribution functions exhibit heavier tails as the arrival rate increases. Moreover, for the queue 2 with slower servers in a group the distribution functions have more heavier tails as the functions for the queue 1 with faster servers.

**Table 7.** $\mathbb{P}[B_{k,max} \le n_k, k \in \{1,2\}]$ as $\lambda$ varies

| $n_k$ | $\lambda = 20$ | $\lambda = 60$ | $\lambda = 80$ | $\lambda = 100$ |
|---|---|---|---|---|
| 0 | 0.99954 | 0.28987 | 0.18176 | 0.13861 |
| 1 | 0.99976 | 0.35109 | 0.18611 | 0.13909 |
| 2 | 0.99998 | 0.41231 | 0.19046 | 0.13956 |
| 3 | 0.99999 | 0.52854 | 0.19990 | 0.14028 |
| 4 | 0.99999 | 0.53782 | 0.20302 | 0.14066 |
| 5 | 0.99999 | 0.72977 | 0.22567 | 0.14203 |
| $\vdots$ | $\vdots$ | $\vdots$ | $\vdots$ | $\vdots$ |
| 10 | 1 | 0.98347 | 0.46075 | 0.15136 |
| $\vdots$ | $\vdots$ | $\vdots$ | $\vdots$ | $\vdots$ |
| 20 | 1 | 0.99998 | 0.98851 | 0.26974 |
| $\vdots$ | $\vdots$ | $\vdots$ | $\vdots$ | $\vdots$ |
| 40 | 1 | 1 | 0.99999 | 0.94821 |

## 6    Conclusion

We have computationally investigated the problem of an optimal allocation of servers in two heterogeneous groups between two parallel queues. We have proposed a comprehensive model accounting waiting costs and usage costs for cross-connectivity. Using dynamic programming approach we have illustrated that the optimal control policy is not a trivial one and strongly depends on the number of customers in each queue. For a fixed control policy we developed algorithms to calculate the mean performance measures and the distribution function for the maximal queue length in a busy period. The obtained results allow us not only to estimate the performance of this queue but also to understand behavioural patterns of the system and sensitivity of system characteristics to the changes of system parameters.

## References

1. Al-Zubaidy, H., Lambadaris, I., Viniotis Y.: Optimal allocation of multiple servers to parallel queues with independent random connectivity. arXiv:1103.1448v3 (2011)
2. Atar, R., Giat, C., Shimkin, N.: The $c\mu/\theta$ rule for many-server queues with abandonment. Oper. Res. **58**(5), 1427–1439 (2010)
3. Ayesta, U., Jacko, P., Novak, V.: Scheduling of multi-class multi-server queueing systems with abandonments. J. Sched. **20**(2), 129–145 (2015). https://doi.org/10.1007/s10951-015-0456-7
4. Buyukkoc, C., Varaiya, P., Walrand, J.: The $c\mu$ rule revisited. Adv. Appl. Probab. **17**(1), 237–238 (1985)

5. Cox, D.R., Smith, W.: Queues. Chapman & Hall, London (1971)
6. Duenyas, I., Oyen, M.P.V.: Stochastic scheduling of parallel queues with set-up costs. Queueing Syst. Theor. Appl. **19**, 421–444 (1995). https://doi.org/10.1007/BF01151932
7. Hofri, M., Ross, K.W.: On the optimal control of two queues with server setup times and its analysis. SIAM J. Comput. **16**(2), 399–420 (1987)
8. Koole, G.: Assigning a single server to inhomogeneous queues with switching costs. Theor. Comput. Sci. **182**(1–2), 203–216 (1997)
9. Liu, Z., Nain, P.: On optimal polling policies. Queueing Syst. Theor. Appl. **11**(1–2), 59–83 (1992)
10. Puterman, M.L.: Markov Decision Processes: Discrete Stochastic Dynamic Programming. Wiley, Hoboken (2005)
11. Rykov, V.V.: Controlled queueing systems (Itogi Nauki i Techniki). In: Probability Theory, Mathematical Statistics, Theoretical Cybernetics, vol. 12, pp. 43–153 (1975). (in Russian)
12. Sennott, L.I.: Average cost optimal stationary policies in infinite state Markov decision processes with unbounded costs. Oper. Res. **37**, 626–633 (1989)
13. Sleptchenko, A., Van Hartem, A., Van Der Heijden, M.: An exact solution for the state probabilities of the multi-class, multi-server queue with preemptive priorities. Queueing Syst. **50**, 81–107 (2005)
14. Van Mieghem, J.A.: Dynamic scheduling with convex delay costs: the generalized $c\mu$ rule. Ann. Appl. Probab. **5**(3), 809–833 (1995)

# Queueing Analysis of Cognitive Radio Networks with Finite Number of Secondary Users

Velika Dragieva[1(✉)] and Tuan Phung-Duc[2]

[1] University of Forestry, 10 Kliment Ohridsky, 1756 Sofia, Bulgaria
dragievav@yahoo.com
[2] University of Tsukuba, 1-1-1 Tennodai, Tsukuba, Ibaraki 305-8573, Japan

**Abstract.** The paper deals with a queueing model, arising in the cognitive radio networks, with one server and two types of users - primary and secondary. The primary users arrive according to a Poisson flow and have preemptive priority over the secondary users, which form a quasi-random input of demands. Each secondary user upon its arrival has to join the orbit regardless of the server state. Service times of both types users follow two distinct arbitrary distributions. We derive recurrence formulas for computing the stationary system state distribution and present numerical results for the main macro-characteristics of the system performance.

## 1 Introduction

Queues with finite-source input (also called closed queues or queues with quasi-random input with intensity $\lambda$) are modeled by assuming that the server/servers serve a finite number of customers (identical units, users, clients), as it is in most of the real situations. Each of these customers generates an attempt for service in interval $(t, t + dt)$ with probability $\lambda dt + o(dt)$ as $dt \to 0$ $(\lambda > 0)$, when the customer is free at time $t$, and 0, when the customer is being served or waiting for service at time $t$, independent of the states of any other customer (unit). Thus, if at time $t$ there are $n$ customers in free state, the probability that in interval $(t, t + dt)$ a call (an attempt for service) arrives is $n\lambda dt + o(dt)$ as $dt \to 0$. These models have been used to analyze the performance of telephone, computer, communication and other systems (see [5,6,8,20,21]).

The characteristic feature of queueing systems with an virtual waiting room, called orbit concerns the behavior of those customers (units, demands, calls) whose service can not start at the moment of their arrival. In the models with an orbit these customers are not lost or allowed to queue, they join the orbit and usually stay in it for an exponentially distributed random time. After this time they either repeat their attempts for service until finding the server idle (in the models with an orbit of retrial customers) or go into free state (in the models with an orbit of blocked customers). Queues with an orbit of retrial or blocked

© Springer Nature Switzerland AG 2020
M. Gribaudo et al. (Eds.): ASMTA 2019, LNCS 12023, pp. 18–32, 2020.
https://doi.org/10.1007/978-3-030-62885-7_2

customers arise in diverse real situations including our daily activity, telephone switching systems, telecommunication and computer networks, call centers, cellular and local area networks, etc. (see [1,4,7,9,27,28]). A systematic account of the fundamental methods and the latest results, as well as an classified bibliography on this topic can be found, for example in [4,16,17,22] and references therein.

Single-server retrial queue with a finite number of customers has been studied in a number of articles by a number of authors: Ohmura and Takahashi [26], Falin and Artalejo [18], Amador [3], Dragieva [10], while Dragieva [11,12] analysed the same model but with an orbit of blocked customers. The corresponding models with retrials, extended with additional features of the service and/or customers behaviour, have been also extensively studied. This includes service with an unreliable server (in [29,31]), service with two phases of the service times (in [30]), service with collisions (in [24]), service with random access (in [19]), service with two-way communication (in [13,14,23,25]). Some of the models with two-way communication can be considered as models with two types of customers. This, for example refers the model analysed in [14] where it is assumed that after some exponentially distributed idle time the server makes outgoing calls, directed to some customers outside the system. A finite-source, single server retrial queue with two types of customers and nonpreemptive priority of the first type customers over the second type is considered by Gómez-Corral in [16].

In the present paper we analyse a single-server queue with two types of customers: primary customers (users, units) which arrive according to a Poison flow, and a finite number of secondary customers which form quasi-random input of demands. Each of the secondary customers have to sense the channel before uses it. Thus, we assume that before accessing to the server the secondary user must go to the orbit. To the best of our knowledge, retrial queues with such behaviour of the users have not been studied before. We also assume that the primary users have preemptive priority over the secondary users.

The motivation for studying such model are the cognitive radio networks, where secondary users opportunistically use the wireless channel (which is granted to primary users) when there are no present primary users. Primary users have preemptive priority over secondary users. In other words, secondary users must evacuate upon the arrival of primary users. Furthermore, before using the channel, secondary users must sense the status of the channel and they can use it only in the idle case. A closely related work is due to Akutsu and Phung-Duc, [2], who consider the same model with Poisson input of secondary users. In [2] the service distributions of primary and secondary users are exponential.

In our analysis we apply the discrete transformations method, common in the investigation of queues with finite source (see [11,12,17,18,21,26,29–31]) and derive recurrent formulas for computing the stationary system state distribution and the main macro-characteristics of the system performance.

Further the paper is organized as follows. In Sect. 2 we describe the model in detail and introduce the necessary notations. Section 3.1 contains formulas for calculation of the stationary joint distribution of the server state and the orbit size. In Sect. 3.2, on the basis of numerical examples we investigate the influence of the system input parameters on the main characteristics of the system performance. In Sect. 4 we consider the convergence of our model to the corresponding queue with Poisson input flow of $SUs$. Section 5 closes the paper and presents some possible further investigations.

## 2   Model Description and Balance Equations

As stated in the Introduction, we consider a queueing model with one server who serves two types of customers (users): primary users ($PU$'s) which arrive according to a Poisson flow with intensity $\lambda_2$, and a finite number of $N$ secondary users ($SUs$). Each of the $SUs$ arrives at the system with rate $\lambda_1$ and has to sense the channel before using it. Thus, we assume that before accessing to the server the $SU$ must go to the orbit for an exponentially distributed time with mean $1/\mu$. After this time, if the server is idle it is immediately occupied by the $SU$, otherwise the $SU$ enters the orbit again and retries for service until finds the server idle.

A $PU$ does not have to sense, its service starts immediately if the server is idle. If the server is busy with a $SU$, this $SU$ is preempted by the newly arrived $PU$ and goes to the orbit from where retries for service after the same exponentially distributed time (with mean $1/\mu$) as the other customers in the orbit. Otherwise, if the server is busy with a $PU$, the newly arrived $PU$ is lost. Service times of the secondary and the primary users follow two distinct arbitrary distributions with distribution functions $B_i(x)$, hazard rate functions $b_i(x) = B_i'(x)/(1 - B_i(x))$, Laplace-Stieltjes transforms $\beta_i(s)$ and means $1/\nu_i$, $i = 1, 2$, respectively. After the service every $SU$ goes to its free state and with intensity $\lambda_1$ can generate a new demand.

The preemptive priority of $PUs$ over $SUs$ allows to consider the model as a single-server, finite-source queue with an orbit and server breakdowns and repairs (or server vacations). The sources (our $SUs$) form a quasi-random input with intensity $\lambda_1$ and every arriving call first goes to the orbit. The server breakdowns after some exponentially distributed life-time with rate $\lambda_2$, regardless of the presence of a source or not. Service and repair times follow two distinct arbitrary distributions with distribution functions $B_i(x)$, $i = 1, 2$, respectively. We would like to note that in such an interpretation of the model, its basic component are the $SUs$, while in cognitive radio networks the primary component are the $PUs$.

Introducing a supplementary variable $z(t)$, equal to the elapsed service time, the state of the system at time $t$ can be described by the Markov process

$$X(t) = \{C(t), R(t), z(t)\},$$

where $C(t)$ denotes the server state at time $t$ (0, 1 or 2, according to the server is idle, busy with a $SU$ or busy with a $PU$ (under repair), and $R(t)$ is the number of $SUs$ in the orbit at time $t$. Because of the finite state space of the Markov process $X(t)$ the stationary regime exists and we can define the limiting probabilities (densities)

$$p_{i,j}(x)dx = \lim_{t \to \infty} P\left(C(t) = i,\, R(t) = j,\, x \le z(t) < x + dx\right),$$

$$p_{i,j} = \lim_{t \to \infty} P\left(C(t) = i, R(t) = j\right) = \int_0^\infty p_{i,j}(x)dx,\ \ i = 1,2,$$

$$p_{0,j} = \lim_{t \to \infty} P\left(C(t) = 0, R(t) = j\right),\ \ \ j = 0, 1, \ldots, N.$$

In the usual way we derive the equations of statistical equilibrium

$$\frac{dp_{1,n}(x)}{dx} = -\left[\lambda_2 + (N - n - 1)\lambda_1 + b_1(x)\right] p_{1,n}(x) + (N - n)\lambda_1 p_{1,n-1}(x), \quad (1)$$

$$p_{1,n}(0) = (n+1)\mu p_{0,n+1}, \tag{2}$$

$n = 0, \ldots, N - 1,$

$$\frac{dp_{2,n}(x)}{dx} = -\left[(N - n)\lambda_1 + b_2(x)\right] p_{2,n}(x) + (N - n + 1)\lambda_1 p_{2,n-1}(x), \tag{3}$$

$$p_{2,n}(0) = \lambda_2\left(p_{0,n} + p_{1,n-1}\right), \tag{4}$$

$$p_{0,n}\left[\lambda_2 + n\mu + (N - n)\lambda_1\right] = p_{0,n-1}\left(N - n + 1\right)\lambda_1 +$$

$$\int_0^\infty p_{1,n}(x)b_1(x)dx + \int_0^\infty p_{2,n}(x)b_2(x)dx, \tag{5}$$

$n = 0, \ldots, N,$ with $p_{0,-1} = p_{1,-1}(x) = p_{2,-1}(x) = p_{1,-1} = 0.$

## 3  Stationary System State Distribution

In this Section, solving the system (1)–(5) we derive recursive formulas for calculating the stationary joint distribution $p_{i,n}$ of the orbit size and the server state and orbit size, and present numerical results for main macro characteristics of the system performance.

### 3.1  Joint Distribution of the Server State and the Orbit Size

Following the method of discrete transformations we write Eqs. (1) and (3) in a matrix form,

$$[\theta_i I_i - A_i]\,\overline{p}_i(x) = 0, \tag{6}$$

$i = 1, 2,$ where

$$\theta_i = b_i(x) + \frac{d}{dx} + \delta_{i,1}\lambda_2,$$

$I_i$ are the identity matrices of order $N - \delta_{i,1}$, $A_i$ are constructed from (1) and (3), respectively in the usual way and $\bar{p}_i(x)$ are the column vectors of the unknown functions $p_{i,j}(x)$,

$$\bar{p}_i(x) = (p_{i,0}(x), \ldots, p_{i,K-2+i}(x))^T .$$

Then we find the matrices $Y_i$ and $\Lambda_i$, such that $Y_i^{-1} A_i Y_i = \Lambda_i$. Here $\Lambda_i$ are diagonal whose diagonal elements are the eigenvalues of $A_i$ with corresponding eigenvectors $\bar{y}_i^{(k)} = \left( y_{i,0}^{(k)}, y_{i,1}^{(k)}, \ldots, y_{i,N-\delta_{i,1}}^{(k)} \right)$ $(k = 0, \ldots, N - \delta_{i,1})$ that form the columns of the matrix $Y_i$. It is not difficult to verify that

$$\Lambda_i = diag\{0, -\lambda_1, \ldots, -(N - \delta_{i,1})\lambda_1\}$$

and that

$$y_{i,n}^{(k)} = \begin{cases} (-1)^{k-(N-n-\delta_{i,1})} \binom{k}{N-n-\delta_{i,1}}, & \text{for } k + n \geq N - \delta_{i,1}, \\ 0, & \text{otherwise,} \end{cases} \tag{7}$$

for all $n = 0, \ldots, N - \delta_{i,1}$, $k = 0, \ldots, N - \delta_{i,1}$, $i = 1, 2$. This means that applying the transformations

$$p_{i,n}(x) = \sum_{k=0}^{N-\delta_{i,1}} y_{i,n}^{(k)} q_{i,k}(x) =$$

$$\sum_{k=N-n-\delta_{i,1}}^{N-\delta_{i,1}} (-1)^{k-(N-n-\delta_{i,1})} \binom{k}{N-n-\delta_{i,1}} q_{i,k}(x), \tag{8}$$

$i = 1, 2$, we can solve (1) and (3). Then, using (5), (2) and (4) we derive formulas for the densities $p_{i,n}(x)$ and the corresponding probabilities $p_{i,n}$ and $p_{0,n}$, $n = 0, \ldots, N - \delta_{i,1}$. These formulas are presented in the next proposition.

**Proposition 1.** *The stationary joint distribution of the server state and the orbit size can be calculated according to the following formulas:*

$$p_{i,n}(x) = e^{-\delta_{i,1}\lambda_2 x} \times$$

$$\sum_{k=N-n-\delta_{i,1}}^{N-\delta_{i,1}} (-1)^{k-(N-n-\delta_{i,1})} \binom{k}{N-n-\delta_{i,1}} \bar{q}_{i,k}(1 - B_i(x))e^{-k\lambda_i x}, \tag{9}$$

$$p_{i,n} = \int_0^\infty p_{i,n}(x)dx =$$

$$\sum_{k=N-n-\delta_{i,1}}^{N-\delta_{i,1}} (-1)^{k-(N-n-\delta_{i,1})} \binom{k}{N-n-\delta_{i,1}} \bar{q}_{i,k} \frac{1 - \bar{\beta}_{i,k}}{k\lambda_i + \delta_{i,1}\lambda_2}, \tag{10}$$

$i = 1, 2, \ n = 0, \ldots, N - \delta_{i,1},$

$$[(N - n)\lambda_1 + \lambda_2 + n\mu] p_{0,n} = p_{0,n-1} (N - n + 1)\lambda_1 +$$

$$\sum_{k=N-n}^{N} (-1)^{k-(N-n)} \binom{k}{N-n} \bar{q}_{2,k} \bar{\beta}_{2,k} +$$

$$(1 - \delta_{n,N}) \sum_{k=N-n-1}^{N-1} (-1)^{k-(N-n-1)} \binom{k}{N-n-1} \bar{q}_{1,k} \bar{\beta}_{1,k}, \tag{11}$$

$n = 0, \ldots, N, \ p_{0,-1} = 0, \ \ where$

$$\bar{\beta}_{i,k} = \beta_i (k\lambda_i + \delta_{i,1}\lambda_2), \quad \frac{1 - \bar{\beta}_{2,k}}{k\lambda_2} = \frac{1}{\nu_2} \ for \ k = 0. \tag{12}$$

*The quantities* $\bar{q}_{1,k}, \bar{q}_{2,k}$ *and* $p_{0,n}$ *are connected by the linear equations*

$$\left[N\lambda_1 + \lambda_2 \left(1 - \bar{\beta}_{2,N}\right)\right] \bar{q}_{2,N} = \lambda_2 \bar{q}_{1,N-1} \bar{\beta}_{1,N-1}, \tag{13}$$

$$\sum_{k=N-n}^{N} (-1)^{k-(N-n)} \binom{k}{N-n} \bar{q}_{2,k} =$$

$$\lambda_2 \sum_{k=N-n}^{N-1} (-1)^{k-(N-n)} \binom{k}{N-n} \left(\frac{1}{n\mu} + \frac{1 - \bar{\beta}_{1,k}}{k\lambda_1 + \lambda_2}\right) \bar{q}_{1,k}, \tag{14}$$

$n = 1, \ldots, N,$

$$\bar{q}_{1,N-n-2} \bar{\beta}_{1,N-n-2} = \sum_{k=N-n-1}^{N} (-1)^{k-(N-n)} \binom{k}{N-n-1} \bar{q}_{2,k} \bar{\beta}_{2,k} -$$

$$p_{0,n} (N - n)\lambda_1 + \sum_{k=N-n-1}^{N-1} (-1)^{k-(N-n-1)} \left[\binom{k}{N-n-2} \bar{\beta}_{1,k} +\right.$$

$$\left.\binom{k}{N-n-1} \left(\frac{(N-n-1)\lambda_1 + \lambda_2}{(n+1)\mu} + 1\right)\right] \bar{q}_{1,k}, \tag{15}$$

*for* $n = 0, \ldots, N - 2,$

$$(\lambda_2 + N\mu) \sum_{k=0}^{N-1} (-1)^k \bar{q}_{1,k} = N\mu \left(p_{0,N-1}\lambda_1 + \sum_{k=0}^{N} (-1)^k \bar{q}_{2,k} \bar{\beta}_{2,k}\right), \tag{16}$$

*for* $n = N - 1.$

*Proof.* As stated above, applying in (1) and (3) transformations (8) for $i = 1, 2$, respectively we get them in the simpler form

$$\frac{dq_{i,n}(x)}{dx} = -\left[n\lambda_i + b_i(x) + \delta_{i,1}\lambda_2\right] q_{i,n}(x),$$

with solutions

$$q_{i,n}(x) = q_{i,n}(0)\left(1 - B_i(x)\right)\exp\left\{-n\lambda_i x\right\}\exp\left\{-\delta_{i,1}\lambda_2 x\right\},$$

$n = 0, \ldots, N - \delta_{i,1}$.

This leads to the following expressions for the densities $p_{i,n}(x)$,

$$p_{i,n}(x) = e^{-\delta_{i,1}\lambda_2 x}\sum_{k=0}^{N-\delta_{i,1}} y_{i,n}^{(k)}\bar{q}_{i,k}(1 - B_i(x))e^{-k\lambda_i x}, \tag{17}$$

where $\bar{q}_{i,k} = q_{i,k}(0)$. Substituting according to these equations in (5) we obtain

$$p_{0,n}\left[\lambda_2 + n\mu + (N - n)\lambda_1\right] = p_{0,n-1}\left(N - n + 1\right)\lambda_1 +$$

$$\sum_{k=0}^{N-1} y_{1,n}^{(k)}\bar{q}_{1,k}\bar{\beta}_{1,k} + \sum_{k=0}^{N} y_{2,n}^{(k)}\bar{q}_{2,k}\bar{\beta}_{2,k}, \tag{18}$$

where $\bar{\beta}_{1,k}$, $\bar{\beta}_{2,k}$ are given by (12). Equations (17) and (18), compared with formulas (7) that define $y_{i,n}^{(k)}$, prove formulas (9)–(11).

Now, using the initial conditions (2) and (4) we derive a system of linear equations for $\bar{q}_{i,k}$. To this end we substitute $p_{0,n}$ from (2) to (4) $(n = 1, \ldots, N)$ which leads to the following relations:

$$p_{2,0}(0) = \lambda_2 p_{0,0},$$

$$p_{2,n}(0) = \lambda_2\left(\frac{p_{1,n-1}(0)}{n\mu} + p_{1,n-1}\right), n = 1, \ldots, N.$$

Substituting here according to (9)–(11) we prove formulas (13)–(14). In a similar way Eqs. (2) lead to another relations between quantities $\bar{q}_{i,k}$ and $p_{0,n}$:

$$\left[(N - n - 1)\lambda_1 + \lambda_2 + (n + 1)\mu\right]\sum_{k=N-n-1}^{N-1}(-1)^{k-(N-n-1)}\binom{k}{N-n-1}\bar{q}_{1,k} =$$

$$(n + 1)\mu\left[p_{0,n}(N - n)\lambda_1 + \sum_{k=N-n-1}^{N}(-1)^{k-(N-n-1)}\binom{k}{N-n-1}\bar{q}_{2,k}\bar{\beta}_{2,k}+\right.$$

$$\left.(1 - \delta_{n,N-1})\sum_{k=N-n-2}^{N-1}(-1)^{k-(N-n-2)}\binom{k}{N-n-2}\bar{q}_{1,k}\bar{\beta}_{1,k}\right],$$

$n = 0, \ldots, N - 1$. Last equations can be presented in the form (15), (16) which ends the proof of the theorem.

Proposition 1 provides a convenient way to calculate the probabilities $p_{i,n}$. We see that all variables $\bar{q}_{i,n}$ and probabilities $p_{i,n}$ are proportional to $\bar{q}_{2,N}$. The coefficients of proportionality of $\bar{q}_{i,n}$ and $p_{0,n}$ can be recursively computed from (11), (13)–(16) by putting $\bar{q}_{2,N} = 1$. To this end we first apply formula (13), then formula (11) for $n = 0$, and calculate the coefficients of $\bar{q}_{1,N-1}$, and $p_{00}$:

$$\bar{q}_{2,N} = 1,$$

$$\bar{q}_{1,N-1} = \frac{\left[N\lambda_1 + \lambda_2\left(1 - \bar{\beta}_{2,N}\right)\right]}{\lambda_2\bar{\beta}_{1,N-1}}\bar{q}_{2,N},$$

$$p_{0,0} = \frac{\bar{q}_{2,N}\bar{\beta}_{2,N} + \bar{q}_{1,N-1}\bar{\beta}_{1,N-1}}{N\lambda_1 + \lambda_2}.$$

Next, after we have calculated the coefficients of proportionality of all quantities $\bar{q}_{1,N-k-1}$, $\bar{q}_{2,N-k}$ and $p_{0,k}$, $k = 0, \ldots, n$, $n = 0, \ldots, N-2$, we can calculate the coefficients of $\bar{q}_{2,N-n-1}$, $\bar{q}_{1,N-n-2}$ and $p_{0,n+1}$. More precisely, from Eq. (14) for $n+1$ we obtain

$$\bar{q}_{2,N-n-1} = \sum_{k=N-n-2}^{N}(-1)^{k-N+n}\binom{k}{N-n-1}\bar{q}_{2,k} -$$

$$\lambda_2\sum_{k=N-n-1}^{N-1}(-1)^{k-N+n}\binom{k}{N-n-1}\left(\frac{1}{(n+1)\mu} + \frac{1 - \bar{\beta}_{1,k}}{k\lambda_1 + \lambda_2}\right)\bar{q}_{1,k},$$

then from (15) we calculate the coefficient of $\bar{q}_{1,N-n-2}$, after which applying (11) for $n+1$ we calculate the coefficient of $p_{0,n+1}$. After this procedure is finished, only the coefficient of $\bar{q}_{2,0}$ and $p_{0,N}$ are still unknown. We calculate $\bar{q}_{2,0}$ from (16) (or from (14) for $n = N$) and then, from (11) for $n = N$, we calculate $p_{0,N}$.

Finally, from (10) we can calculate the coefficients of $p_{i,n}$ ($i = 1,2$, $n = 0, \ldots, N - \delta_{i,1}$) which allows to calculate $\bar{q}_{2,N}$ using the normalizing condition

$$P_0 + P_1 + P_2 = 1,$$

where $P_i$, ($i = 0,1,2$) is the stationary server state distribution,

$$P_i = \lim_{t\to\infty}P\left(C(t) = i\right).$$

## 3.2   Main Macro-characteristics of the System Performance

The purpose of this Section is to investigate the relations between some of the system input parameters (intensity of the $SUs$, $\lambda_1$, intensity of the $PUs$, $\lambda_2$, intensity of the retrials, $\mu$) and the main macro-characteristics of the system performance:

- server state distribution, $P_i, i = 0, 1, 2$;
- mean orbit size,

$$E[R] = \lim_{t \to \infty} E[R(t)] = E_0 + E_1 + E_2,$$

$$E_i = \sum_{n=0}^{N-\delta_{i,1}} np_{i,n}, i = 0, 1, 2;$$

- mean rate of generation of $SUs$,

$$\Lambda = (N - E[R] - P_1)\lambda_1;$$

- mean waiting time of a $SU$ in the orbit,

$$E[W] = \Lambda^{-1}E[R];$$

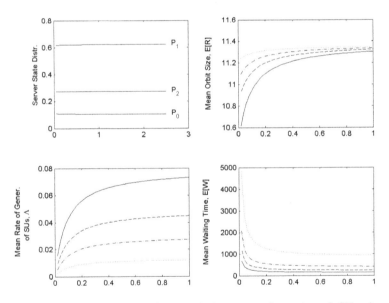

**Fig. 1.** Basic performance macro-characteristics versus intensity of $SUs$, $\lambda_1$. ($\lambda_2 = 0.3, \mu = 0.2, \nu_1 = 0.1, \nu_2 = 0.8, N = 12$)

To illustrate the influence of the system input parameters on the main macro-characteristics of the system performance we present numerical results, calculated for four different distributions of the service times, both for $SU's$ and $PU's$, with means, $1/\nu_i$ ($i = 1, 2$):

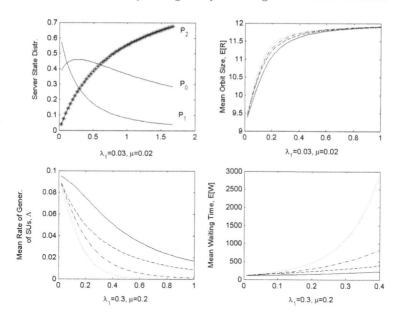

**Fig. 2.** Basic performance macro-characteristics versus intensity of $PUs$, $\lambda_2$. ($\nu_1 = 0.1, \nu_2 = 0.8, N = 12$)

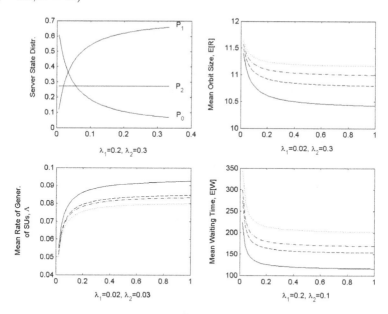

**Fig. 3.** Basic performance macro-characteristics versus intensity of retrials, $\mu$. ($\nu_1 = 0.1, \nu_2 = 0.8, N = 12$)

- Deterministic distributions, equal to $1/\nu_i$, respectively, presented with dotted lines;
- Erlang distributions with parameters 4 and $4\nu_i$, respectively, presented with dash-dot lines;
- Exponential distributions with parameters $\nu_i$, respectively, presented with solid lines;
- Uniform distributions in the intervals $(0, 2/\nu_i)$, respectively, presented with dashed lines.

On Fig. 1 we see the influence of the intensity of $SUs$, $\lambda_1$ on the server state distribution, $P_i$ $(i = 0, 1, 2)$ (the upper-left corner), on the mean orbit size, $E[R]$ (the upper-right corner), on the mean rate of generation of $SU's$, $\Lambda$ (the lower-left corner), and on the mean waiting time, $E[W]$ (the lower-right corner). Figures 2 and 3 have the same structure as Fig. 1 but show the dependance of the same macro-characteristics on the intensity of $PUs$, $\lambda_2$ and on the retrial rate, $\mu$, respectively. In order to present all probabilities $P_i$ $(i = 0, 1, 2)$ on the same figure (Fig. 1–3, upper-left corners) we show only the results for exponentially distributed service times. If all four distributions are included, a 12-graph picture will be produced. Because of the preemptive priority of $PUs$ over $SUs$, the probability $P_2$ that the server is busy with a $PU$ should coincide with the probability of a busy server in the standard $M/G/1/1$ queue. Thus, for the results presented on the upper-left corners of all figures, $P_2$ should be equal to $\lambda_2/(\nu_2 + \lambda_2)$. On Fig. 1 and Fig. 3 we see that, as it should be expected, the intensity of $SU's$, $\lambda_1$, and the retrial intensity, $\mu$ do not affect $P_2$ whose value is exactly equal to $\lambda_2/(\nu_2 + \lambda_2)$. Fig. 2 shows the coincidence between the values of $P_2$ and the corresponding values $\lambda_2/(\nu_2 + \lambda_2)$, depicted by a line of stars.

Our numerical results show that the intensity of $SUs$, $\lambda_1$ has almost no influence on the server state distribution $P_i$ $(i = 0, 1, 2)$ (Fig. 1) which can be intuitively expected if all $SUs$ first go to the orbit. On the same figure we see that for small values of $\lambda_1$ the mean orbit size is an increasing function of this parameter, while for large values of $\lambda_1$ it remains constant between $N - 1$ and $N$. This is also expected: large values of the $SU$ intensity will keep all of them, with exception of the user under service, in the orbit. The influence of the $PUs$ intensity, $\lambda_2$ on the mean orbit size, $E[R]$ is similar (Fig. 2): $E[R]$ is first an increasing function of $\lambda_2$ and then, for larger values of this parameter all $N$ $SUs$ are in the orbit. Large values of the retrial rate, $\mu$ also do not affect the values of the mean orbit size (it is a constant), while for small values of $\mu$ the mean orbit size is a decreasing function of this parameter (Fig. 3).

The behaviour of the mean rate of generation of $SUs$, $\Lambda$ can be easily explained by the formula for $\Lambda$ calculation and the presented values of $P_1$ and $E[R]$. The same refers for the behaviour of the mean waiting time, $E[W]$.

Comparing our results with those for the single-server queues with quasi-random input and "standard" retrials (when the call joins the orbit only if its service can not start at the time of arrival) we find a considerable difference. All papers dealing with such queues show that the mean waiting time and, in some models, the mean rate of generation of incoming customers first increase as

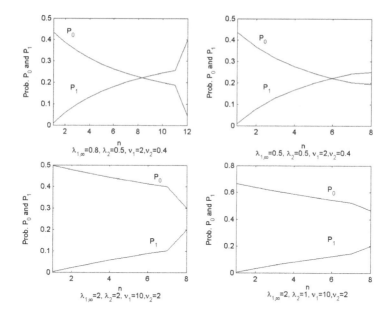

**Fig. 4.** Probabilities of idle server, $P_0$ and of server busy with a $SU$, $P_1$ versus number of $SUs$, $N$. ($\mu = 0.01, N = 5 + 50(n - 1), \lambda_1 = \lambda_{1,\infty}/N$)

functions of the arriving intensity, then decrease, i.e. have a point of maximum [13–15, 18, 29–31]. The assumption that all $SUs$ join the orbit regardless of the server state, as it is in cognitive radio networks, strongly changes these unexpected and theoretically unproven properties of the finite-source retrial queues. We see that the behavior of mean waiting time and of mean rate of generation of incoming calls is clear and intuitively expected. Strange and unexpected is the behaviour of the idle server probability, $P_0$ (Fig. 2) which first increases with the increase of $PUs$ intensity, $\lambda_2$ then decreases.

From all presented results we see that the distribution of the service times has a strong influence on the considered relations, which depends on the particular relation and on the values of the input parameters.

## 4    Comparison with the Model with Poisson Input Flow of $SUs$

In this Section we consider some properties of the stationary server sate distribution when the number $N$ of all $SUs$ converges to infinity, the arrival intensity of these users, $\lambda_1$ converges to 0, in such a way that $N\lambda_1$ converges to some constant, which will be denoted as $\lambda_{1,\infty}$. Under these conditions our model with finite number of $SUs$ should converge to the corresponding model in which the $SUs$ arrive according to a Poisson flow with intensity $\lambda_{1,\infty}$. As stated in the Introduction, this model is studied by Akutsu and Phung-Duc [2] in the case of exponential service times. They obtain the condition for existing a stationary state of the considered Markov process,

$$\frac{\lambda_{1,\infty}}{\nu_1} < \frac{\nu_2}{\nu_2 + \lambda_2},$$

and formulas for computing the stationary server state distribution, $P_{i,\infty}$ ($i = 0, 1, 2$),

$$P_{0,\infty} = \frac{\nu_2}{\nu_2 + \lambda_2} - \frac{\lambda_{1,\infty}}{\nu_1}, \quad P_{1,\infty} = \frac{\lambda_{1,\infty}}{\nu_1}, \quad P_{2,\infty} = \frac{\lambda_2}{\nu_2 + \lambda_2}. \tag{19}$$

On Fig. 4 we present the values of $P_0$ and $P_1$ in the model with $N$ $SUs$. The last values, shown at the right side of each graph are equal to the values of $P_{i,\infty}$ ($i = 0, 1$), calculated according to formulas (19). The values of $P_2$ are not presented since we saw in previous Section that they coincide with the values of $P_{2,\infty}$.

The results shown on Fig. 4 illustrate the convergence of probabilities $P_i$ to the corresponding probabilities $P_{i,\infty}$ ($i = 0, 1$), and the influence of the system input parameters on the rate of this convergence. Decreasing $\lambda_{1,\infty}$ from 0.8 to 0.5 at the same values of the other parameters accelerates the convergence, and even for $N = 305$ both probabilities $P_0$ and $P_1$ reach the values of $P_{0,\infty}$ and $P_{1,\infty}$, respectively (Fig. 4, upper two graphs). Similarly, the lower two graphs of Fig. 4 show that the smaller values of $PUs$ intensity, $\lambda_2$ improve the convergence of $P_0$ and $P_1$ to $P_{0,\infty}$ and $P_{1,\infty}$.

## 5   Concluding Remarks

In this paper we analyze a queueing system with one server, that serves two types of users, called secondary and primary users, respectively. The primary users arrive according to a Poisson flow, while the secondary users form a quasi-random input of demands. The service times of both users follow two distinct arbitrary distributions. Before accessing the server, each secondary user has to sense the server, because of which it is assumed that upon arrival it goes to the orbit regardless of the server state and tries to get service after an exponentially distributed random time. Finally, the primary users have preemptive priority over the secondary ones which allows to consider the model as a finite-source queue with server breakdowns and repairs. The breakdowns can occur during the busy and the idle periods of the server. This model is appropriate for modeling the cognitive radio networks.

We derive recurrence formulas for computing the stationary distribution of the system states and the main macro-characteristics of the system performance. On the basis of numerical examples we investigate some relations between these characteristics and the system input parameters. The obtained formulas can be applied for further investigation of this model, like the busy period properties, the waiting time process, including the number of retrials that a customer in the orbit will make before entering service, and others.

# References

1. Aguir, S., Karaesmen, E., Aksin, O., Chauvet, F.: The impact of retrials on call center performance. OR Spectr. **26**, 353–376 (2004)
2. Akutsu, K., Phung-Duc, T.: Analysis of retrial queues for cognitive wireless networks with sensing time of secondary users. In: Phung-Duc, T., Kasahara, S., Wittevrongel, S. (eds.) QTNA 2019. LNCS, vol. 11688, pp. 77–91. Springer, Cham (2019). https://doi.org/10.1007/978-3-030-27181-7_6
3. Amador, J.: On the distribution of the successful and blocked events in retrial queues with finite number of sources. In: Proceedings of the 5th International Conference on Queueing Theory and Network Applications, pp. 15–22 (2010)
4. Artalejo, J., Gómez-Corral, A.: Retrial Queueing Systems: A Computational Approach. Springer, Heidelberg (2008). https://doi.org/10.1007/978-3-540-78725-9
5. Balazsfalvi, G., Sztrik, J.: A tool for modeling distributed protocols. PIK **31**(1), 39–44 (2008)
6. Biro, J., Bérczes, T., Kőrösi, A., Heszberger, Z., Sztrik, J.: Discriminatory processor sharing from optimization point of view. In: Dudin, A., De Turck, K. (eds.) ASMTA 2013. LNCS, vol. 7984, pp. 67–80. Springer, Heidelberg (2013). https://doi.org/10.1007/978-3-642-39408-9_6
7. Choi, B., Shin, Y.W., Ahn, W.C.: Retrial queues with collision arising from unslotted CSMA/CD protocol. Queueing Syst. **11**(4), 335–356 (1992)
8. Cooper, R.: Introduction to Queueing Theory, 2nd edn. Edward Arnold, London (1981)
9. Deslauriers, A., L'Ecuyer, P., Pichitlamken, J., Ingolfsson, A., Avramidis, A.: Markov chain models of a telephone call center with call blending. Comput. Oper. Res. **34**, 1616–1645 (2007)
10. Dragieva, V.: A finite source retrial queue: number of retrials. Commun. Stat. Theor. Meth. **42**(5), 812–829 (2013)
11. Dragieva, V.: Number of lost calls during the busy period in an M/G/1//N queue with inactive orbit. In: Dudin, A., Nazarov, A., Yakupov, R. (eds.) ITMM 2015. CCIS, vol. 564, pp. 85–98. Springer, Cham (2015). https://doi.org/10.1007/978-3-319-25861-4_8
12. Dragieva, V.I.: Steady state analysis of the M/G/1//N queue with orbit of blocked customers. Ann. Oper. Res. **247**(1), 121–140 (2015). https://doi.org/10.1007/s10479-015-2025-z
13. Dragieva, V., Phung-Duc, T.: Two-way communication M/M/1//N retrial queue. In: Thomas, N., Forshaw, M. (eds.) ASMTA 2017. LNCS, vol. 10378, pp. 81–94. Springer, Cham (2017). https://doi.org/10.1007/978-3-319-61428-1_6
14. Dragieva, V., Phung-Duc, T.: An M/G/1//K retrial queue with outgoing calls. In: Proceedings of the 13th International Conference on Queueing Theory and Network Application, QTNA 2018, pp. 9–13, Tsukuba, Japan, 25–27 July 2018 (2018)
15. Gómez-Corral, A.: Analysis of a single-server retrial queue with quasi-random input and nonpreemptive priority. Comput. Math. Appl. **43**, 767–782 (2002)
16. Gómez-Corral, A., Phung-Duc, T.: Retrial queues and related models. Ann. Oper. Res. **247**(1), 1–2 (2016). https://doi.org/10.1007/s10479-016-2305-2
17. Falin, G., Templeton, J.: Retrial Queues. Chapman and Hall, London (1997)
18. Falin, G., Artalejo, J.: A finite source retrial queue. Eur. J. Oper. Res. **108**, 409–424 (1998)

19. Fiems, D., Phung-Duc, T.: Light-traffic analysis of random access systems without collisions. Ann. Oper. Res. **277**(2), 311–327 (2017). https://doi.org/10.1007/s10479-017-2636-7
20. Jain, R.: The Art of Computer Systems Performance Analysis. Wiley, New York (1991)
21. Jaiswal, N.: Priority Queues. Academic Press, New York (1969)
22. Kim, J., Kim, B.: A survey of retrial queueing systems. Ann. Oper. Res. **247**(1), 3–36 (2015). https://doi.org/10.1007/s10479-015-2038-7
23. Kuki, A., Sztrik, J., Tóth, Á., Bérczes, T.: A contribution to modeling two-way communication with retrial queueing systems. In: Dudin, A., Nazarov, A., Moiseev, A. (eds.) ITMM 2018, WRQ 2018. Communications in Computer and Information Science, vol. 912, pp. 236–247. Springer, Cham (2018). https://doi.org/10.1007/978-3-319-97595-5_19
24. Nazarov, A., Sztrik, J., Kvach, A.: Some features of a finite-source M/GI/1 retrial queueing system with collisions of customers. In: Proceedings of the 20th International Conference, DCCN, pp. 186–200 (2017)
25. Nazarov, A., Sztrik, J., Kvach, A.: Asymptotic sojourn time analysis of finite-source M/M/1 retrial queuing system with two-way communication. In: Dudin, A., Nazarov, A., Moiseev, A. (eds.) ITMM 2018, WRQ 2018. CCIS, vol. 912, pp. 172–183. Springer, Cham (2018). https://doi.org/10.1007/978-3-319-97595-5_14
26. Ohmura, H., Takahashi, Y.: An analysis of repeated call model with a finite number of sources. Electron. Commun. Jpn. **68**, 112–121 (1985)
27. Tran-Gia, P., Mandjes, M.: Modeling of customer retrial phenomenon in cellular mobile networks. IEEE J. Sel. Areas Commun. **15**, 1406–1414 (1997)
28. Van Do, T., Wochner, P., Berches, T., Sztrik, J.: A new finite-source queueing model for mobile cellular networks applying spectrum renting. Asia-Pac. J. Oper. Res. **31**, 14400004 (2014)
29. Wang, J., Zhao, L., Zhang, F.: Analysis of the finite source retrial queues with server breakdowns and repairs. J. Ind. Manage. Optim. **7**(3), 655–676 (2011)
30. Wang, J., Wang, F., Sztrick, J., Kuki, A.: Finite source retrial queue with two phase service. Int. J. Oper. Res. **3**(4), 421–440 (2017)
31. Zhang, F., Wang, J.: Performance analysis of the retrial queues with finite number of sources and service interruption. J. Korean Stat. Soc. **42**, 117–131 (2013)

# On Reliability of a Double Redundant Renewable System

Vladimir Rykov[1,2]([⊠]) [ID]

[1] Gubkin Russian State University, 65 Leninsky Prospect,
Moscow 119991, Russian Federation
vladimir_rykov@mail.ru
[2] Peoples' Friendship University of Russia (RUDN University),
6 Miklukho-Maklaya St., Moscow 117198, Russian Federation

**Abstract.** The reliability of a double redundant renewable system with generally distributed life- and repair times of its components is considered. For its investigations the theory of decomposable regenerative processes is used. Reliability function and stationary characteristics of the system with the help of a special integral transformation are founded in closed form.

**Keywords:** Reliability · Generally distributed life- and repair times · Reliability function · Steady state probabilities

## 1 Introduction

Investigations of reliability of systems with arbitrary distributions of their components life and repair times are interesting both from theoretical and practical points of view. From theoretical point of view these investigations serve to introduction and development of new mathematical methods. From practical point of view they create the base for reliability analysis of different complex systems with generally distributed life- and repair times distributions of their components.

An investigation of asymptotic behaviour of a double cold redundant renewable system under quick restoration has been proposed by Gnedenko and Soloivjev [1–3]. The investigation of analogous system when one of the input distributions (life- or repair times) has an exponential distribution has been done in [4,5]. Then investigations of more complex systems including those with heterogeneous components and using the simulation methods has been proposed in [6,7]. An application of these models one can find in [8]. The review of these investigations and some of their generalization one can find in [9]. Study of the systems with dependent failures of their components was done in [10].

The publication has been prepared with the support of the RUDN University Program 5-100 (The Sects. 1 and 2). The Sects. 3, 4 and 5 was funded by RFBR, project number 20-01-00575A.

However most of these investigations based on the exponential property one of the input distributions (life- or repair time).

In this paper the investigation of a homogeneous cold double redundant system with generally distributed both life- and repair times distributions is considered. Due to modified Laplace transform, the closed form representations of the reliability function and the steady state probabilities (s.s.p.'s) of the system has been obtained. The investigation is based on the stochastic equations for considered in the paper random variables (r.v.'s) and the theory of regenerative processes due to Smith [11].

The material is organized as follows. In the next section the problem set, some assumptions, and notations are introduced. In the Sect. 3 the moment generation functions (m.g.f.'s) of the regeneration period (the system life cycle) and times to the first and between system failures are calculated in terms of modified Laplace transforms of their components' life- and repair time distributions. In Sects. 4 and 5 the system s.s.p.'s are calculated. The paper ends with conclusion.

## 2    The Problem Set. Assumptions and Notations

Consider an homogeneous cold double redundant renewable system with generally distributed life- and repair times, which will be denoted as $<GI_2/GI/1>$. The system consists from two components, each of which can be in two states: operational, denoted by 0 and failed, denoted by 1. Denote by $A_i$ $(i = 1, 2, \dots)$ and by $B_i$ $(i = 1, 2, \dots)$ the sequences of its components life- and repair times. It is assumed that the successive lifetimes of the same and different components as well as all their repair times are independent and identically distributed. Denote also by $A(t) = \mathbf{P}\{A_i \leq t\}$ and $B(t) = \mathbf{P}\{B_i \leq t\}$ their cumulative distribution functions (c.d.f.) and suppose that the instantaneous failures and repairs are impossible and their mean times are finite:

$$A(0) = B(0) = 0, \quad a = \int_0^\infty (1 - A(x))dx < \infty, \quad b = \int_0^\infty (1 - B(x))dx < \infty.$$

The system fails when both its components fail. Denote by $E = \{i = 0, 1, 2\}$ the system set of states, where $i$ means the number of failed components and introduce a random process $J = \{J(t), t \geq 0\}$, where

$$J(t) = \text{number of failed components in time } t.$$

Denote by $T$ time to the system failures after its repair. We are interesting in calculation of the system reliability function

$$R(t) = \mathbf{P}\{T > t\},$$

the time dependent

$$\pi_j(t) = \mathbf{P}\{J(t) = j\},$$

and the system s.s.p.'s

$$\pi_j = \lim_{t \to \infty} \pi_j(t) \equiv \lim_{t \to \infty} \mathbf{P}\{J(t) = j\},$$

as well as the availability coefficient

$$K_{\text{av.}} = \pi_0 + \pi_1.$$

Further the following notations will be used. The m.g.f.'s of r.v.'s $A_i$ and $B_i$ are denoted by

$$\tilde{a}(s) = \int_0^\infty e^{-sx} dA(x), \quad \tilde{b}(s) = \int_0^\infty e^{-sx} dB(x),$$

We also introduce modified Laplace transform and call them by *modified m.g.f.*—(m.m.g.f.) by the formulas:

$$\tilde{a}_B(s) = \int_0^\infty e^{-sx} B(x) dA(x), \quad \tilde{b}_A(s) = \int_0^\infty e^{-sx} A(x) dB(x), \tag{1}$$

as well as their derivatives with sign minus in point zero, to which we will refer as *modified mean values*

$$a_B = -\frac{d}{ds}\tilde{a}_B(s)\Big|_{s=0} = \int_0^\infty x B(x) dA(x),$$

$$b_A = -\frac{d}{ds}\tilde{b}_A(s)\Big|_{s=0} = \int_0^\infty x A(x) dB(x). \tag{2}$$

The probabilities $\mathbf{P}\{B \le A\}$ and $\mathbf{P}\{B \ge A\}$ connected with these transformations by the expressions

$$\tilde{a}_B(0) = \int_0^\infty B(x) dA(x) = \mathbf{P}\{B \le A\} \equiv p,$$

$$\tilde{b}_A(0) = \int_0^\infty A(x) dB(x) = \mathbf{P}\{B > A\} \equiv q = 1 - p.$$

Note some properties of transformations $\tilde{a}_B(s)$, $\tilde{b}_A(s)$:

$$\tilde{a}_B(s) + \tilde{a}_{1-B}(s) = \tilde{a}(s), \quad \tilde{b}_A(s) + \tilde{b}_{1-A}(s) = \tilde{b}(s)$$

$$\tilde{a}_B(s) + \tilde{b}_A(s) = s \int_0^\infty e^{-sx} A(x) B(x) dx$$

## 3   Reliability Function

Process $J$ is a regenerative one with delay and in the Fig. 1 its trajectory is presented. The regeneration times of the process $J$ are:

$$S_0 = 0, \ S_1 = A_1, \ S_2 = S_1 + G_1, \ \ldots, \ S_{k+1} = S_k + G_k, \ \ldots,$$

where regeneration periods $G_k$ $(k = 1, 2, \ldots)$ are the time intervals between two successive returns of the process $J$ into the state 1 after the system failure and repair, when one of the system's components begins operate and the other begins to be repaired. This r.v. satisfies to the equation

$$G = \begin{cases} A + G & \text{if } B < A, \\ B & \text{if } B > A. \end{cases} \tag{3}$$

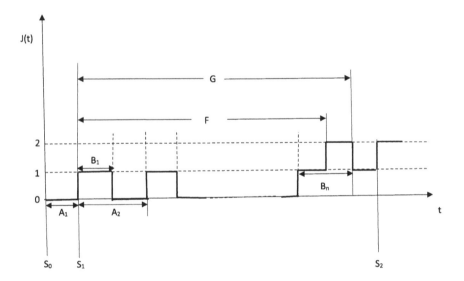

**Fig. 1.** Trajectory of the process $J$.

Denote by $F$ the time between system failures (the time to the failure after the system failure and repair) and by $F_1$ the time to the first system failure. From the Fig. 1 one can see that the time between the system failures $F$ satisfies to the stochastic equation

$$F = \begin{cases} A + F & \text{if } B < A, \\ A & \text{if } B > A, \end{cases} \tag{4}$$

and the time to the first system failure equals to:

$$F_1 = A + F. \tag{5}$$

The c.d.f.'s of these r.v.'s denote by

$$G(t) = \mathbf{P}\{G \le t\}, \quad F(t) = \mathbf{P}\{F \le t\}, \quad F_1(t) = \mathbf{P}\{F_1 \le t\}$$

and their m.g.f.'s by $\tilde{g}(s)$, $\tilde{f}(s)$, $\tilde{f}_1(s)$.

Applying Laplace transform to stochastic Eq. (3) one can find the equation

$$\tilde{g}(s) = \mathbf{E}\left[e^{-sG}\right] = \int_0^\infty e^{-st} dG(t)$$

$$= \int_0^\infty dA(x) \left[B(x)e^{-sx}\tilde{g}(s) + \int_{y>x} e^{-sy} dB(y)\right] =$$

$$= \tilde{g}(s) \int_0^\infty e^{-sx} B(x) dA(x) + \int_0^\infty e^{-sy} A(y) dB(y) =$$

$$= \tilde{g}(s)\tilde{a}_B(s) + \tilde{b}_A(s),$$

from which the expression for the regeneration period m.g.f. follows:

$$\tilde{g}(s) = \mathbf{E}[e^{-sG}] = \frac{\tilde{b}_A(s)}{1 - \tilde{a}_B(s)}. \tag{6}$$

Analogously from the equality (4) one can find

$$\tilde{f}(s) = \mathbf{E}\left[e^{-sT}\right] = \int_0^\infty e^{-st} dF(t) =$$

$$= \int_0^\infty e^{-sx} \left[B(x)\tilde{f}(s) + (1 - B(x))\right] dA(x) =$$

$$= \tilde{f}(s)\tilde{a}_B(s) + \tilde{a}_{1-B}(s),$$

from which the expression for m.g.f. of the system times between failures follows:

$$\tilde{f}(s) = \frac{\tilde{a}_{1-B}(s)}{1 - \tilde{a}_B(s)}. \tag{7}$$

In these equations the notations (1, 2) are used. At least using the equality (5) the following representation for m.g.f. of the system time to the first failure holds:

$$\tilde{f}_1(s) = \frac{\tilde{a}(s)\tilde{a}_{1-B}(s)}{1 - \tilde{a}_B(s)}. \tag{8}$$

From the expression (6) one can find the expectations of the regeneration period $\mathbf{E}[G]$,

$$g = \mathbf{E}[G] = -\tilde{g}'(0) = \frac{a_B + b_A}{q}, \tag{9}$$

Taking into account that the Laplace transform of any c.d.f. connected with its Laplace-Stiltjes transform (m.g.f. of appropriate r.v.) by the correlation

$$\tilde{F}(s) = \int_0^\infty e^{-st} F(t)dt = \frac{1}{s} \int_0^\infty e^{-st} dF(t) = \frac{1}{s}\tilde{f}(s)$$

for Laplace transform $\tilde{R}(s)$ of the reliability function $R(t) = 1 - F(t)$ from (7) using the properties of m.m.g.f. one can find

$$\tilde{R}(s) = \frac{1}{s} - \tilde{F}(s) = \frac{1}{s}(1 - \tilde{f}(s)) = \frac{1 - \tilde{a}_B(s) - \tilde{a}_{1-B}(s)}{s(1 - \tilde{a}_B(s))} = \frac{1 - \tilde{a}(s)}{s(1 - \tilde{a}_B(s))}. \quad (10)$$

At that for the mean system life time it follows:

$$\mathbf{E}[F] = \tilde{R}(0) = \frac{a}{1 - \tilde{a}_B(0)} = \frac{a}{q}. \quad (11)$$

## 4   System Steady State Probabilities

For the system (and the process $J$) state probabilities calculation we use the renewal theory, accordingly to which the state probabilities of the process in any time $t$

$$\pi_j(t) = \mathbf{P}\{J(t) = j\} \quad (j = 0, 1, 2)$$

can be represented in terms of its distribution at separate regeneration period

$$\pi_j^{(1)}(t) = \mathbf{P}\{J(t) = j, \ t < G\} \quad (j = 0, 1, 2)$$

and the renewal function $H(t)$ as follows

$$\pi_i(t) = \pi_i^{(1,1)}(t) + \int_0^\infty dH(u)\pi_j^{(1)}(t - u). \quad (12)$$

In our case as it was mentioned before the process $J$ is a regenerative process with delay (see Fig. 1). Thus the the first period differs from the others, has the duration $A$ and if the system functioning begins when both components are in good state the process $J$ distribution on it has the form

$$\pi_j^{(1,1)}(t) = \mathbf{P}\{J(t) = j, \ t < A\} = \delta_{j0}(1 - A(t)), \quad (13)$$

In terms of the c.d.f. $G(t) = \mathbf{P}\{G_i \le t\}$ of r.v.'s $G_i$ the renewal function $H(t)$ is determined as follows

$$H(t) = \sum_{k \ge 1} \mathbf{P}\left\{\left[\sum_{1 \le i \le k} G_i\right] \le t\right\} = \sum_{k \ge 1} G^{*k}(t), \quad (14)$$

Accordingly to the key Smith theorem for regenerative processes [11] the s.s.p.'s of the process exists and has the form

$$\lim_{t\to\infty} \pi_j(t) = \frac{1}{\mathbf{E}[G]} \int_0^\infty \pi_j^{(1)}(t)dt = \frac{1}{\mathbf{E}[G]} \int_0^\infty \mathbf{E}[1_{\{J(t)=j,\ t\leq G\}}]dt = \frac{\mathbf{E}[G_j]}{\mathbf{E}[G]}, \quad (15)$$

where $G_j$ is the time spent by the system in its state $j$, $(j \in E)$ during regeneration time $G$. For calculation these values and their expectations consider the process $J$ behavior on a separate regeneration period.

## 5  Investigation of the System Behavior on a Separate Regeneration Period

Consider the system behavior on a separate regeneration period. Remind that $G$ denoted the length of regeneration period, and $G_j$ is the time spent by the process $J$ in its states $j = \{0, 1, 2\}$ on its. As it can be seen from the Fig. 1 they satisfies to the following stochastic equations:

$$G_0 = \begin{cases} 0 & \text{if } A \leq B, \\ A - B + G_0 & \text{if } A > B; \end{cases}$$

$$G_1 = \begin{cases} A & \text{if } A \leq B, \\ B + G_1 & \text{if } A > B; \end{cases}$$

$$G_2 = \begin{cases} B - A & \text{if } A \leq B, \\ G_2 & \text{if } A > B. \end{cases} \quad (16)$$

Calculating the expectations of r.v.'s $G_i$ $(i = 0, 1, 2,)$ form (16) one can find the equations:

$$g_0 \equiv \mathbf{E}[G_0] = \int_{x\geq 0} \int_{y\leq x} (x - y + g_0)dA(x)dB(y) =$$

$$= pg_0 + \int_{x\geq 0} [xB(x) - \int_{y\leq x} ydB(y)]dA(x) = pg_0 + a_B - b + b_a,$$

$$g_1 \equiv \mathbf{E}[G_1] = \int_{x\geq 0} x(1 - A(x))dB(x) + \int_{x\geq 0} (x + g_1)B(x))dA(x) =$$

$$= a + b - (a_B + b_A) + g_1p,$$

$$g_2 \equiv \mathbf{E}[G_2] = \int_{x\geq 0} \int_{y\leq x} (x - y)dA(y)dB(x) + g_2p =$$

$$= pg_2 + \int_{x\geq 0} [xA(x) - \int_{y\geq x} ydA(y)]dB(x) =$$

$$= pg_2 + b_A - \int_{y\geq 0} ydA(y)(1 - B(y)) = pg_2 + b_A - a + A_B.$$

Thus it holds

$$g_0 = \frac{a_b + b_A - b}{q}, \quad g_1 = \frac{(a+b) - (a_b + b_A)}{q}, \quad g_2 = \frac{a_B + b_A - a}{q},$$

and therefore taking into account the expression (9) for $\mathbf{E}[\mathbf{G}]$ the stationary probabilities are:

$$\pi_0 = 1 - \frac{b}{a_B + b_A} \quad \pi_1 = \frac{a+b}{a_B + b_A} - 1, \quad \pi_2 = 1 - \frac{a}{a_B + b_A}. \qquad (17)$$

## 6   Conclusion

With the help of special transformations the closed form representation of the reliability function and the steady state probabilities for the double redundant renewable system with generally distributed life- and repair times of its components are found.

**Acknowledgments.** The publication has been prepared under support of the RUDN University Program "5-100" (Sects. 1 and 2) and funded by RFBR according to the research projects No. 17-01-00633 (Sects. 3 and 4) and No. 17-07-00142 (Sect. 5).

## References

1. Gnedenko B.V.: On cold double redundant system. Izv. AN SSSR. Texn. Cybern. **4**, 3–12 (1964). (in Russian)
2. Gnedenko B.V.: On cold double redundant system with restoration. Izv. AN SSSR. Texn. Cybern. **5**, 111–118 (1964). (in Russian)
3. Solov'ev A.D.: On reservation with quick restoration. Izv. AN SSSR. Texn. Cybern. **1**, 56–71 (1970). (in Russian)
4. Rykov, V., Ngia, T.A.: On sensitivity of systems reliability characteristics to the shape of their elements life and repair time distributions. Vestnik PFUR. Ser. Mathematics. Inform. Phys. **3**, 65–77 (2014). (in Russian)
5. Efrosinin, D., Rykov, V.: Sensitivity analysis of reliability characteristics to the shape of the life and repair time distributions. In: Dudin, A., Nazarov, A., Yakupov, R., Gortsev, A. (eds.) ITMM 2014. CCIS, vol. 487, pp. 101–112. Springer, Cham (2014). https://doi.org/10.1007/978-3-319-13671-4_13
6. Rykov V., Kozyrev D., Zaripova E.: Modeling and simulation of reliability function of a homogeneous hot double redundant repairable system. In: Proceedings 31st European Conference on Modeling and Simulation, ECMS 2017, pp. 701–705 (2017). https://doi.org/10.7148/2017-0701
7. Rykov, V., Kozyrev, D.: On sensitivity of steady-state probabilities of a cold redundant system to the shapes of life and repair time distributions of its elements. In: Pilz, J., Rasch, D., Melas, V.B., Moder, K. (eds.) IWS 2015. SPMS, vol. 231, pp. 391–402. Springer, Cham (2018). https://doi.org/10.1007/978-3-319-76035-3_28
8. Efrosinin D., Rykov V., Vishnevskiy V.: Sensitivity of reliability models to the shape of life and repair time distributions. In: Proceedings of the 9th International Conference on Availability, Reliability and Security, ARES 2014, pp. 430–437. IEEE (2014). Published in CD: 978-I-4799-4223-7/14. https://doi.org/10.1109/ARES2014.65

9. Rykov, V.: On Reliability of renewable systems. In: Vonta, I., Ram, M. (eds.) Reliability Engineering. Theory and Applications, pp. 173–196. CRC Press (2018)
10. Kozyrev, D., Kolev, N., Rykov, V.: Reliability function of renewable system under Marshall-Olkin failure model. Reliab. Theor. Appl. **13**, 39–46 (2018)
11. Smith, W.L.: Regenerative stochastic processes. Proc. R. Soc. Ser. A **232**, 6–31 (1955)

# Queueing Analysis for a Mixed Model of Carsharing and Ridesharing

Ayane Nakamura[1], Tuan Phung-Duc[2(✉)], and Hiroyasu Ando[2]

[1] College of Policy and Planning Sciences, University of Tsukuba,
Tsukuba, Ibaraki 305-8577, Japan
s1611294@u.tsukuba.ac.jp
[2] Faculty of Engineering, Information and Systems, University of Tsukuba, Tsukuba,
Ibaraki 305-8577, Japan
{tuan,ando}@sk.tsukuba.ac.jp

**Abstract.** Sharing services such as carsharing and ridesharing have been widely spreading in our modern society. This paper considers "Car Ride Sharing (CRS)" which is a mixed model of carsharing and ridesharing. CRS is a new means of transportation which shares "car" and "ride" simultaneously and it is expected to reduce the redistribution cost of cars. We introduce CRS to a scenario with two points between which transportation service by bus already exists. We model this system as a queuing model, formulated by a Markov chain of GI/M/1-type, and derive the steady state distribution of the queue-length. Besides, we derive the stability condition which must be satisfied to ensure proper function of CRS and some performance measures such as average waiting time of customers. Based on these results, we conduct some numerical experiments to show the effectiveness and possibility of CRS.

**Keywords:** Sharing economy · Mobility as a Service (MaaS) · Queueing analysis

## 1 Introduction

Sharing economy is a system where assets or services are shared between private individuals. This is particularly utilized among young people, and it is said that we have reached turning point in times from "possession" to "sharing" [1]. The market size of the conventional rental industry is predicted to increase 1.4 times from 2013 to 2025, but surprisingly, that of sharing economy is expected to expand by 22 times. It is clear that the latter will catch up with the former soon [1].

The Japanese government is promoting the Society 5.0 aiming at a super smart society with the harmony of cyber and physical spaces using Artificial Intelligence (AI), big data and IoT [2]. One of the goals of the Society 5.0 is MaaS (Mobility as a Service) which optimally integrates self-driving and various mobility services [3]. The sharing economy has been spreading to these mobilities, for example, "carsharing" and "ridesharing". Carsharing is a system that service

© Springer Nature Switzerland AG 2020
M. Gribaudo et al. (Eds.): ASMTA 2019, LNCS 12023, pp. 42–56, 2020.
https://doi.org/10.1007/978-3-030-62885-7_4

providers lend cars to users such as Zipcar in America. On the other hand, ridesharing is a system that customers ride on a car with other customers, Uber is one of the most famous examples. In carsharing, customers have to drive by themselves, but they can set the destination and the time of use as they like. On the other hand, customers can utilize ridesharing if they can find other customers who have same destinations or directions with them. Therefore, we have to make a mechanism which allows users to choose a service according to their personal situations. In addition, we have to discuss the possibility of the coexistence of those services with existing transportation.

There are two types of carsharing. The first one is "round trip type" which means customers return cars to the points of the delivery, the second one is "one way type" which means customers place cars at their destinations. The former has an advantage that it is easy for service providers to manage cars, but it takes time and effort for customers to return cars. The latter is convenient for customers, but it may cause uneven distribution of cars and takes cost to redistribute cars. To solve this, the research on redistribution of cars and that on optimization of the boarding system were conducted [4,5]. For ridesharing, the research on optimization of drivers and riders matching, and that on the possibility for introduction of autonomous cars were conducted [6,7].

It is very meaningful to think about the service which reduces the drawbacks while taking advantages of the conventional sharing services. Then, we suggest Car Ride Sharing (CRS) as a new means of transportation in this study. CRS is considered to outcome various shortcomings of carsharing. This is a new service defined as a mixed model of carsharing and ridesharing. In a previous research, simulation experiments showed the effectiveness of CRS [8]. In this paper, we model and analyze this new service system using queueing theory, and conduct some numerical experiments to discuss its feature and effectiveness.

The structure of this paper is organized as follows. In Sect. 2, we state the detailed mechanism of CRS. In Sect. 3, we present the analysis of the model and we derive the stability conditions and some performance measures. Section 4 shows some numerical results for performance measures. In Sect. 5, we present concluding remarks.

## 2   The CRS System

The CRS proposed in this paper is defined as a new means of transportation that performes carsharing and ridesharing simultaneously. In this section, we describe this CRS system specifically.

### 2.1   Premise

In this paper, we model a situation in which two means of transportation, "bus" and "CRS" coexist between two points, a station and a university (bus stop), as shown in Fig. 1. In the morning, people who came to university by their private cars provide these cars as "Car Ride Sharing Car (car)". We assume that the minimum and the maximum numbers of passengers for a car are $m$ and $n$, respectively. Customers can use these cars to move between the university and the station, but it is necessary that these cars are returned within a certain period of time before the owners come home.

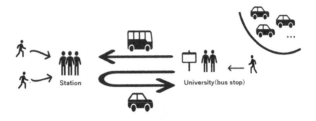

**Fig. 1.** Image of Car Ride Sharing (CRS).

## 2.2    Station Demand

At the station, customers who want to use CRS gather and make the demand. We assume that in the station, there are two states; 1 (with a demand) and 0 (without a demand). The state 1 can be interpreted that a certain number of customers gathered. We assume that the demands occur according to a Poisson process with rate $\delta$ and the presence or absence of the demand in the station is instantaneously transmitted to customers at the university through a website or an application etc. We do not consider the possibility of multiple demands occurring at the same time, cancellation of the demand due to the arrival of other transportation means such as a bus or personal reasons of customers.

## 2.3    Movement from the University to the Station

There are two means of transportation from the university to the station, bus and car. It is assumed that buses depart from the university to the station according to a Poisson process with rate $\mu$. We further assume that the capacity of a bus is $l$ and the university is the first bus stop, i. e., a bus is empty upon arrival. People lined up at the university continuously get on a bus on a first-come, first-served basis within the limit of the capacity. On the other hand, CRS occurs from the university at intervals following the exponential distribution with parameter $\sigma$ under the conditions that "There is a demand at the station" and "There are $m$ or more people at the university". In the same way as buses, people get on a car on a first-come, first-served basis within the limit of the capacity. Here, we assume that all the customers at the university have a driving license, and there is an agreement that they can go to the station either by CRS or bus.

## 2.4   Takeover of the Car at the Station and Car Return

A car departing at the university goes to the station, and after arriving at the station, those who are in the car get off. Then, people waiting at the station take over the car and start CRS to the university and the demand disappears. By this series of flows, it can be considered to meet the demand on the station side.

After the car arrives at the university, people who got on at the station get off and the car is returned to its original position. By this, the demand for people who want to go to the station from the university and the opposite ones are satisfied simultaneously under the condition that the car is returned to the university within a certain time (in other words, no uneven distribution of cars occurs). Therefore, CRS can be considered as a service model that overcomes various shortcomings of carsharing.

# 3   Modelling

In this section, we model CRS described in the previous section using queueing theory.

## 3.1   Parameters Used in the Model

In the beginning, we summarize the parameters in the model as follows (Table 1).

**Table 1.** The parameters used in the model.

| Parameters | Definitions |
|---|---|
| $\lambda$ | The arrival rate of customers at the university |
| $\mu$ | The departure rate of buses |
| $\sigma$ | The rate of the occurrence of CRS |
| $\delta$ | The rate of the occurrence of the demand |
| $l$ | The capacity of buses |
| $m$ | The minimum number of passengers of cars |
| $n$ | The maximum number of passengers of cars |

## 3.2   Markov Chain and Stability Condition

The CRS system described in the previous section can be formulated using a GI/M/1-type Markov chain. Let $S(t)$ and $N(t)$ denote the number of the demand at the station and the number of customers waiting at the university. It is easy to see that $\{(S(t), N(t)); t \geq 0\}$ forms a Markov chain in the state space $S = \{(i, j); i = 0, 1, j = 0, 1, 2 \ldots\}$. Our Markov chain is of GI/M/1-type, where $N(t)$ is the level and $S(t)$ is the phase. For example, see Fig. 2 for the transitions among states for $m = 2$, $n = 4$, and $l = 30$. Assuming that the Markov chain is stable, we define steady state probabilities as follows.

$$\pi_{i,j} = \lim_{t \to \infty} P(S(t) = i, N(t) = j).$$

The infinitesimal generator $Q$ of our Markov chain is given as follows.

$$Q =$$

| | 0 | 1 | 2 | ... | m-1 | m | m+1 | ... | n | n+1 | n+2 | ... | l-n | l-n+1 | ... | l | l+1 | l+2 | ... |
|---|---|---|---|---|---|---|---|---|---|---|---|---|---|---|---|---|---|---|---|
| 0 | $B_{0,0}$ | $A_0$ | $O$ | ... | $O$ | $O$ | $O$ | ... | $O$ | $O$ | $O$ | ... | $O$ | $O$ | ... | $O$ | $O$ | $O$ | ... |
| 1 | $C_{-l}$ | $B_{1,1}$ | $A_0$ | | $O$ | $O$ | $O$ | ... | $O$ | $O$ | $O$ | ... | $O$ | $O$ | ... | $O$ | $O$ | $O$ | ... |
| 2 | $C_{-l}$ | $O$ | $B_{2,2}$ | | $O$ | $O$ | $O$ | ... | $O$ | $O$ | $O$ | ... | $O$ | $O$ | ... | $O$ | $O$ | $O$ | ... |
| ⋮ | ⋮ | ⋮ | | | | | | | ⋮ | ⋮ | ⋮ | ⋮ | | ⋮ | | ⋮ | ⋮ | ⋮ | ... |
| m-1 | $C_{-l}$ | $O$ | $O$ | | $B_{m-1,m-1}$ | $A_0$ | $O$ | ... | $O$ | $O$ | $O$ | ... | $O$ | $O$ | ... | $O$ | $O$ | $O$ | ... |
| m | $B_{m,0}$ | $O$ | $O$ | ... | $O$ | $C_0$ | $A_0$ | | $O$ | $O$ | $O$ | ... | $O$ | $O$ | ... | $O$ | $O$ | $O$ | ... |
| m+1 | $B_{m+1,0}$ | $O$ | $O$ | ... | $O$ | $O$ | $C_0$ | | $O$ | $O$ | $O$ | ... | $O$ | $O$ | ... | $O$ | $O$ | $O$ | ... |
| ⋮ | ⋮ | ⋮ | | | | | | | | | | | | | | | | | |
| n | $B_{n,0}$ | $O$ | $O$ | ... | $O$ | $O$ | $O$ | | $C_0$ | $A_0$ | $O$ | ... | $O$ | $O$ | ... | $O$ | $O$ | $O$ | ... |
| n+1 | $C_{-l}$ | $C_{-n}$ | $O$ | ... | $O$ | $O$ | $O$ | ... | $O$ | $C_0$ | $A_0$ | | $O$ | $O$ | ... | $O$ | $O$ | $O$ | ... |
| n+2 | $C_{-l}$ | $O$ | $C_{-n}$ | | $O$ | $O$ | $O$ | ... | $O$ | $O$ | $C_0$ | | | | | | | | ... |
| ⋮ | ⋮ | ⋮ | | | | | | | | | | | | | | | | | ... |
| l | $C_{-l}$ | $O$ | $O$ | ... | $O$ | $O$ | $O$ | ... | $O$ | $O$ | $O$ | | $C_{-n}$ | $O$ | | $C_0$ | $A_0$ | $O$ | ... |
| l+1 | $O$ | $C_{-l}$ | $O$ | ... | $O$ | $O$ | $O$ | ... | $O$ | $O$ | $O$ | ... | $O$ | $C_{-n}$ | | $O$ | $C_0$ | $A_0$ | |
| l+2 | $O$ | $O$ | $C_{-l}$ | ... | | | | ... | | | | ... | | | ... | | | $C_0$ | |
| ⋮ | ⋮ | ⋮ | ⋮ | | | | ... | | | | | | | ... | | | | |

where,

$$A_0 = \begin{pmatrix} \lambda & 0 \\ 0 & \lambda \end{pmatrix},$$

$$B_{0,0} = \begin{pmatrix} -(\lambda + \delta) & \delta \\ 0 & -\lambda \end{pmatrix},$$

$$B_{1,1} = B_{2,2} = \ldots = B_{m-1,m-1} = \begin{pmatrix} -(\lambda + \mu + \delta) & \delta \\ 0 & -(\lambda + \mu) \end{pmatrix},$$

$$B_{m,0} = B_{m+1,0} = \ldots = B_{n,0} = \begin{pmatrix} \mu & 0 \\ \sigma & \mu \end{pmatrix},$$

$$C_0 = \begin{pmatrix} -(\lambda + \mu + \delta) & \delta \\ 0 & -(\lambda + \mu + \sigma) \end{pmatrix},$$

$$C_{-n} = \begin{pmatrix} 0 & 0 \\ \sigma & 0 \end{pmatrix},$$

$$C_{-l} = \begin{pmatrix} \mu & 0 \\ 0 & \mu \end{pmatrix},$$

and $O$ is the zero matrix with an appropriate dimension.

Defining $\pi_j = (\pi_{0,j}, \pi_{1,j})$ and $\pi = (\pi_0, \pi_1, \pi_2, \ldots)$, the equilibrium equations are given as follows.

$$\pi Q = \mathbf{0}, \tag{1}$$

where $\mathbf{0}$ is a vector of zero with an appropriate size. Furthermore, from the probability normalization, we have,

$$\pi e = 1, \tag{2}$$

where $e$ is a vertical vector of an appropriate size with all elements 1.

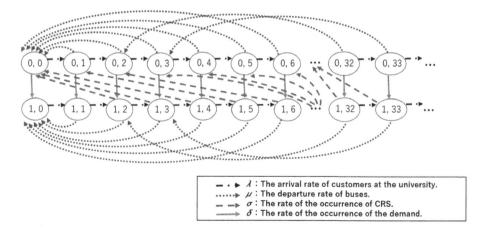

**Fig. 2.** State transition diagram and parameters corresponding to each arrow ($m = 2$, $n = 4$, and $l = 30$).

Because our Markov chain is homogeneous in level (level-independent) from level $m$ ($N(t)$), the steady state vectors $\pi_j$ ($j \geq m - 1$) have a geometric form:

$$\pi_{k+1} = \pi_k R, \qquad k \geq m - 1,$$

and thus

$$\pi_j = \pi_{m-1} R^{j-m+1}, \qquad j \geq m - 1.$$

Here, $R$ is minimal non-negative solution of the following equation.

$$A_0 + RC_0 + R^{n+1}C_{-n} + R^{l+1}C_{-l} = \mathbf{0}.$$

$R$ can be numerically obtained using the following iterative equation started with $R = O$.

$$R = -(A_0 + R^{n+1}C_{-n} + R^{l+1}C_{-l})C_0^{-1}.$$

By rewriting Eq. (1), the unknown quantities $\pi_0, \ldots, \pi_{m-1}$ are determined by the following set of equations.

$$\pi_0 B_{0,0} + \pi_1 C_{-l} + \pi_2 C_{-l} + \ldots + \pi_{m-1} C_{-l} + \pi_{m-1} R B_{m,0} + \pi_{m-1} R^2 B_{m+1,0}$$
$$+ \ldots + \pi_{m-1} R^{n-m+1} B_{n,0} + \pi_{m-1} R^{n-m+2} C_{-l}$$
$$+ \pi_{m-1} R^{n-m+3} C_{-l} + \ldots + \pi_{m-1} R^{l-m+1} C_{-l} = \mathbf{0},$$

$$\tag{3}$$

$$\pi_k A_0 + \pi_{k+1} B_{k+1,k+1} + \pi_{m-1} R^{k+n-m+2} C_{-n} + \pi_{m-1} R^{k+l-m+2} C_{-l} = \mathbf{0}$$

$$\tag{4}$$

$$(0 \leq k \leq m - 2).$$

Also, we obtain Eq. (5) by writing down Eq. (2) using $R$.

$$1 = \sum_{k=1}^{m-1} \pi_{k-1} \mathbf{e} + \pi_{m-1}(I - R)^{-1}\mathbf{e}, \tag{5}$$

By solving Eq. (3) and Eq. (5), $\pi_0, \ldots, \pi_{m-1}$ can be obtained, and the steady state probability is determined.

Next, we derive the stability condition, which is the condition for the existence of the steady state probability. First, we define the infinitesimal generator $A$ as follows.

$$A = A_0 + C_0 + C_{-n} + C_{-l} = \begin{pmatrix} -\delta & \delta \\ \sigma & -\sigma \end{pmatrix}.$$

Assuming that the steady state probability of the phase (with or without the demand at the station) is $\eta = (\eta_1, \eta_2)$, we have

$$\eta A = \mathbf{0}, \tag{6}$$

$$\eta \mathbf{e} = 1. \tag{7}$$

By solving Eq. (6) and Eq. (7), we obtain

$$\eta_1 = \frac{\sigma}{\sigma + \delta}, \qquad \eta_2 = \frac{\delta}{\sigma + \delta}.$$

Here, the rate which the level (the number of customers waiting at the university) decreases by $n$ is $\eta C_{-n}\mathbf{e}$, the rate which the level decreases by $l$ is $\eta C_{-l}\mathbf{e}$, and the rate which the level increases by 1 is $\eta A_0\mathbf{e}$. Therefore, the stability condition is expressed by Eq. (8) (See e.g., [9]).

$$n\eta C_{-n}\mathbf{e} + l\eta C_{-l}\mathbf{e} > \eta A_0\mathbf{e},$$

$$\iff \lambda < \frac{n\sigma\delta}{\sigma + \delta} + l\mu. \tag{8}$$

The stability condition indicates that the mean number of customers departing from the university per unit time is greater than that arrived at the bus stop per unit time, and it is interpreted that the number of customers waiting at the university does not increase explosively if this condition is satisfied.

### 3.3   Performance Measures

In this section, we derive some performance measures for our model. The mean number of customers waiting at the university $E[L]$ is given by

$$E[L] = \sum_{i=0}^{\infty} i\pi_i \mathbf{e}$$

$$= \sum_{i=0}^{m-2} i\pi_i \mathbf{e} + \sum_{i=m-1}^{\infty} i\pi_i \mathbf{e}$$

$$= \sum_{i=0}^{m-2} i\pi_i \mathbf{e} + \sum_{i=0}^{\infty} (m-1+i)\pi_{m-1} R^i \mathbf{e}$$

$$= \sum_{i=0}^{m-2} i\pi_i \mathbf{e} + \pi_{m-1}\{(m-1)(I-R)^{-1} + (I-R)^{-2}R\}\mathbf{e}$$

$$= \sum_{i=0}^{m-2} i\pi_i \mathbf{e} + \pi_{m-1}(I-R)^{-1}\{(m-1)I + (I-R)^{-1}R\}\mathbf{e}.$$

Using Little's law, the mean waiting time for customers at the university $E[W]$ is derived as follows.

$$E[W] = \frac{E[L]}{\lambda}.$$

Finally, we obtain the mean number of CRS occurrences per unit time $E[C]$ as follows.

$$E[C] = \sum_{i=m}^{\infty} \sigma\pi_{1,i}$$

$$= \sum_{i=1}^{\infty} \sigma\pi_{m-1} R^{i-1} \begin{pmatrix} 0 \\ 1 \end{pmatrix} - \sigma\pi_{m-1} \begin{pmatrix} 0 \\ 1 \end{pmatrix}$$

$$= \sigma\pi_{m-1}(I-R)^{-1} \begin{pmatrix} 0 \\ 1 \end{pmatrix} - \sigma\pi_{m-1} \begin{pmatrix} 0 \\ 1 \end{pmatrix}.$$

## 4   Numerical Examples

In this section, we conduct some numerical experiments on performance measures defined in Sect. 3.3.

### 4.1   Mean Waiting Time for Customers

First, we present the results of the numerical experiments for the mean waiting time of customers at the university when we fix $\mu = 8$, $m = 2$, $n = 4$, and $l = 30$, and change the value of $\lambda$, $\sigma$ and $\delta$. Figure 3 and Fig. 4 show $E[W]$ against $\lambda$ for various $\sigma$ and $\delta$. The effectiveness of CRS can be confirmed by the result that $E[W]$ decreases as $\sigma$ and $\delta$ increase because it becomes easier to carry out CRS. Also, when either $\sigma$ or $\delta$ becomes 0 (that is, no CRS is performed at all), $E[W]$ increases monotonously, but is convex if both $\sigma$ and $\delta$ are positive. This means

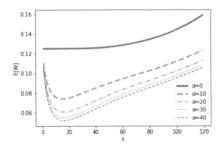
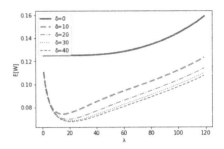

**Fig. 3.** $E[W]$ versus $\lambda$ ($\delta = 10$).     **Fig. 4.** $E[W]$ versus $\lambda$ ($\sigma = 10$).

that no matter how large $\sigma$ or $\delta$ is, the conditions for the occurrence of CRS are not satisfied if there are too few people at the bus stop ($\lambda$ is too small), and as a result, the waiting time is not too short.

The results suggest that it is possible to set $\lambda$ which minimizes $E[W]$ for various parameters. For example, we show $\lambda$ that minimizes $E[W]$ for various $\sigma$ where we fix $\delta = 10, 20, 30$, and 40 in Fig. 5 and $\lambda$ that minimizes $E[W]$ for various $\delta$ where we fix $\sigma = 10, 20, 30$, and 40 in Fig. 6. If $\sigma$ and $\delta$ can be predicted and $\lambda$ can be controlled, $E[W]$ can be minimized by adopting the optimal value of $\lambda$ in these figures. As Fig. 5 and Fig. 6 show, $\lambda$ which minimizes $E[W]$ increases as $\sigma$ and $\delta$ increase. It may be a natural result because more customers at the university are needed in the situation where CRS occurrence rate and the demand are high. It is also proved that $\lambda$ tends to converge to a constant value as the value of the horizontal axis increases in both Fig. 5 and Fig. 6. This means that the mean number of the occurrences of CRS does not change so much if only one of $\sigma$ or $\delta$ increases, so $\lambda$ that minimizes $E[W]$ does not change either.

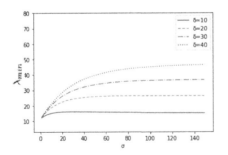
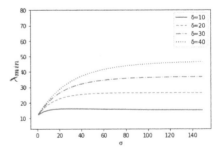

**Fig. 5.** $\lambda$ to minimize $E[W]$ versus $\sigma$ ($\delta = 10, 20, 30$, and 40).     **Fig. 6.** $\lambda$ to minimize $E[W]$ versus $\delta$ ($\sigma = 10, 20, 30$, and 40).

## 4.2  Capacity of a Bus

Next, we consider $E[W]$ in the case of changing $l$ from 20 to 60 in the step size of 10. We fix $\mu = 8$, $m = 2$, and $n = 4$ and conduct some numerical experiments. Figure 7 shows $E[W]$ for each $l$ when $\sigma = 10$ and $\delta = 10$, and Fig. 8 shows $E[W]$ for each $l$ when $\sigma = 15$ and $\delta = 15$.

As shown in Fig. 7 and Fig. 8, $E[W]$ is insensitive to $l$ until $\lambda$ reaches a certain value while $E[W]$ decreases with $l$ when $\lambda$ exceeds that value. Therefore, in terms of the waiting time, it is suggested that the larger the capacity of a bus, the better it is, but it is meaningless to make it too large. Also, comparing Fig. 7 and Fig. 8, it is indicated that $E[W]$ is smaller for the same $l$ in Fig. 8 where more CRS is performed. These observations are in line with intuition.

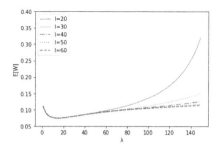 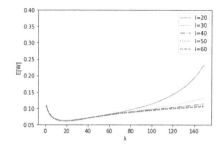

**Fig. 7.** $E[W]$ versus $\lambda$ ($\sigma = 10, \delta = 10$).    **Fig. 8.** $E[W]$ versus $\lambda$ ($\sigma = 15, \delta = 15$).

Then, we discuss the optimal $l$ for each $\lambda$. Generally, it is assumed that it costs money to increase the size of the capacity. Based on this, we assume that it is the best for service providers to introduce a bus with a smaller capacity if the mean of the waiting time of customers does not change. In other words, we assume that service providers want to minimize $E[W]$ with the smallest value of $l$. Figure 9 and Fig. 10 plot the optimal $l$ based on this assumption for the parameters set in Fig. 7 and Fig. 8, respectively. We increase the value of $l$ until the difference of $E[W]$ is less than or equal to $e^{-10}$ and obtain the optimal value of $l$.

In this way, it is possible to find the optimal $l$ for a given $\lambda$ and to apply this result in the real world. It is proved that the larger $\lambda$ is, the larger the optimal $l$ is, and the optimal $l$ becomes smaller under the condition where CRS is frequently performed. Also, this time we assume that service providers prioritize $E[W]$ to simplify the discussion, but it is possible to make flexible decision-making that is more responsive to various situations by defining how much they emphasize the financial cost by using a cost function.

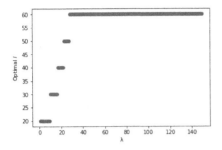

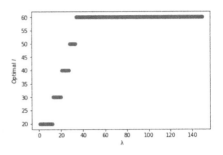

**Fig. 9.** Optimal $l$ versus $\lambda$ ($\sigma = 10, \delta = 10$).

**Fig. 10.** Optimal $l$ versus $\lambda$ ($\sigma = 15, \delta = 15$).

### 4.3   Minimum Number of Passengers of a Car

In this section, we consider $E[W]$ when the minimum number of passengers of a car $m$ is changed from 1 to 5 in the case $\mu = 8$, $n = 6$, and $l = 30$. Figure 11 shows $E[W]$ for various $m$ with $\sigma = 10$ and $\delta = 10$, and Fig. 12 shows $E[W]$ for various $m$ with $\sigma = 15$ and $\delta = 15$. Overall, for any fixed $m$, $E[W]$ is smaller in Fig. 12 where CRS is performed more often than in Fig. 11. This is a reasonable result. In addition, it is interesting to see that $E[W]$ increases monotonically when $m = 1$, but has a down ward convex shape when $m$ is more than one in both the figures. This means that when $m = 1$, the conditions of occurrence of CRS are satisfied at the moment when one or more people arrive at the university, as long as there is a demand at the station. However, if $m$ is more than one, it is difficult to meet the condition of "there are $m$ or more people in the university", which is one of the conditions for the occurrence of CRS. In other words, regardless the value of $\sigma$ and $\delta$, CRS rarely occurs with small $\lambda$.

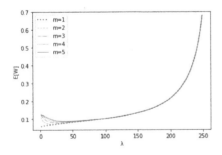

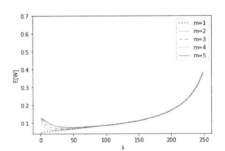

**Fig. 11.** $E[W]$ versus $m$ ($\sigma = 10, \delta = 10$).

**Fig. 12.** $E[W]$ versus $m$ ($\sigma = 15, \delta = 15$).

Next, let us look at the mean number of times that CRS occurs per unit time $E[C]$ for various $m$. Figure 13 and Fig. 14 show $E[C]$ for the parameters set in Fig. 11 and Fig. 12, respectively. Overall, the smaller $m$ is, the larger $E[C]$ is, and there is almost no difference in $E[C]$ when $\lambda$ is extremely small or $\lambda$ increases near the boundary of the stability condition. The results are reasonable because the smaller $m$ is, the easier it is to meet the conditions for the occurrence of CRS.

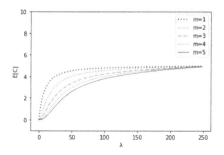

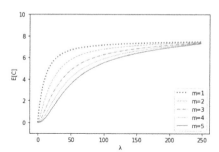

**Fig. 13.** $E[C]$ versus $m$ ($\sigma = 10, \delta = 10$).

**Fig. 14.** $E[C]$ versus $m$ ($\sigma = 15, \delta = 15$).

From the above results, we consider the optimal $m$. It is expected that financial costs such as gasoline and maintenance costs increase as CRS occurs, that is, as $E[C]$ increases. As the consideration of a bus capacity $l$ in the previous section, we discuss the optimal $m$ by giving top priority to the waiting time of customers. Based on this assumption, it is preferable to set smaller $m$ so that $E[W]$ becomes smaller if $\lambda$ is not too large. However, it is considered to be optimal to set larger $m$ in order to minimize $E[C]$ if $\lambda$ is more than a certain value (about 100 or more in this parameter setting) where there is almost no difference in $E[W]$. Figure 15 shows the optimal $m$ when $\sigma = 10$ and $\delta = 10$, and Fig. 16 shows the optimal $m$ when $\sigma = 15$ and $\delta = 15$. It is indicated that the optimal $m$ tends to increase as $\lambda$ increases. Here, we obtain the optimal $m$ by increasing it until the difference in $E[W]$ is less than or equal to 0.001.

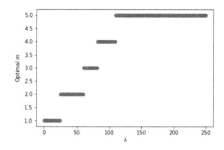

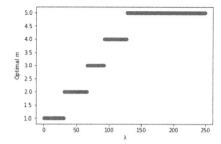

**Fig. 15.** Optimal $m$ versus $\lambda$ ($\sigma = 10, \delta = 10$).

**Fig. 16.** Optimal $m$ versus $\lambda$ ($\sigma = 15, \delta = 15$).

## 4.4 Maximum Number of Passengers of a Car

Finally, we consider $E[W]$ for various $n$ which is the maximum number of passengers of a car. We conduct some numerical experiments with $\mu = 8$, $m = 2$, and $l = 30$. Figure 17 shows $E[W]$ for various $n$ when $\sigma = 10$ and $\delta = 10$, and Fig. 18 shows $E[W]$ for various $n$ when $\sigma = 15$ and $\delta = 15$. Both of the graphs become convex downward because $m$ is more than one and $E[W]$ in Fig. 18 is smaller than that in Fig. 17 because CRS occurs more often in the latter than in the former. Besides, it is indicated that $E[W]$ decreases as $n$ increases. The larger the value of $n$, the more customers who can go to the station with one CRS, so this is an intuitive result.

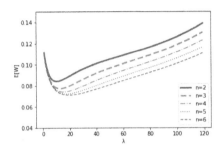

**Fig. 17.** $E[W]$ versus $n$ ($\sigma = 10, \delta = 10$).

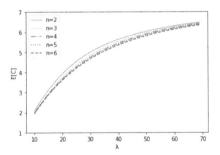

**Fig. 18.** $E[W]$ versus $n$ ($\sigma = 15, \delta = 15$).

Furthermore, we state the results of the numerical experiments on $E[C]$ with various $n$. Figure 19 and Fig. 20 show $E[C]$ against $\lambda$ with the parameters set in Fig. 17 and Fig. 18, respectively. In each case, $E[C]$ decreases slightly as $n$ increases, which is intuitively convincing. From the above results, the mean of the waiting time for customers decreases and the mean number of times that CRS occurs per unit time decreases (that is, the cost for CRS is reduced) as $n$ becomes large. Thus, it is considered that it is better for both of customers and

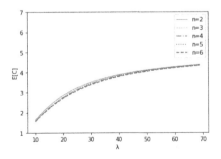

**Fig. 19.** $E[C]$ versus $n$ ($\sigma = 10, \delta = 10$).

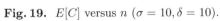

**Fig. 20.** $E[C]$ versus $n$ ($\sigma = 15, \delta = 15$).

service providers to increase $n$ as long as the physical space of a car is allowed. In other words, it turns out that there are no incentives to set $n$ smaller than the actual size of a car, which is a natural result.

## 5   Conclusion

In this paper, we have modeled and analyzed CRS, which is a new means of transportation that performs carsharing and ridesharing simultaneously using queueing theory. According to the results of the numerical experiments, we have shown that the average waiting time at the university decreases as the demand and the occurrence rate of CRS increases. Furthermore, by performing some numerical experiments by changing a bus capacity and the minimum and the maximum number of customers in a car, we have also considered these optimal values in various situations.

As the future issue, the improvement of the model is mentioned. In the current model, only 0 or 1 demand can be considered at the same point in time. It is clear that we have to incorporate multiple demands at the same time in our model. Also, we should consider the arrival of buses and customers following the general distributions, and the psychological aspect such as whether customers want to use CRS or not. Moreover, we would like to work on modeling a traffic system between multiple points. This is possible to bring the model closer to the real situation, and to enable useful discussions and recommendations in the real world by making such improvements. The model with multiple points is presumed to be a new queueing network, so it is also considered to be a very challenging task from a theoretical point of view.

It is also necessary to have a discussion of prices. Even if the waiting time of customers is reduced, it would be meaningless if there were people who make financial loss compared to before the introduction of CRS. In addition, for the perspective of car providers, the participations of the service will be merely volunteer unless they can get benefits above a certain level. Therefore, constructing a market design that brings incentives to both customers and car providers is also an important issue in the future.

**Acknowledgments.** This study was supported by the joint program of the University of Tsukuba and Toyota Motor Corporation, titled "Research on the next generation social systems and mobilities".

## References

1. Sumitomo Mitsui Banking Corporation Homepage. https://www.smbc.co.jp/hojin/report/investigationlecture/resources/pdf/3_00_CRSDReport039.pdf. Accessed 10 July 2019. (in Japanese)
2. Cabinet Office of Japan Homepage. https://www8.cao.go.jp/cstp/kihonkeikaku/index5.html. Accessed 10 July 2019. (in Japanese)
3. Prime Minister's Office of Japan Homepage. https://www.kantei.go.jp/jp/singi/keizaisaisei/pdf/miraitousi2018zentai.pdf. Accessed 10 July 2019. (in Japanese)

4. Uesugi, K., Mukai, N., Watanabe, T.: Allocation method in car sharing system. http://www.ieice.org/iss/de/DEWS/DEWS2007/pdf/d2-1.pdf. Accessed 10 July 2019. (in Japanese)
5. Correia, G., Antunes, A.: Optimization approach to depot location and trip selection in one-way carsharing systems. Transp. Res. Part E **47**, 233–247 (2012)
6. Agatz, N., Erera, A., Savelsbergh, M., Wang, X.: Optimization for dynamic ride-sharing: a review. Eur. J. Oper. Res. **223**, 295–303 (2012)
7. Katsuki, H., Azuma, K., Taniguchi, M.: Introduction possibility of shared-type self-driving cars to intersuburban transportation-from the viewpoint of space-time characteristics and personal attributes of trips. J. City Plann. Inst. Japan **52**(3), 769–775 (2017). (in Japanese)
8. Ando, H., Takahara, I., Osawa, Y.: Mobility services in university campus. Commun. Oper. Res. Soc. Japan **64**(8), 447–452 (2019). (in Japanese)
9. Takine, T.: Beyond M/M/1 - invitation to the quasi-birth-death process -. Commun. Oper. Res. Soc. Japan **59**(4), 3–8 (2014). (in Japanese)

# An All Geometric Discrete-Time Multiserver Queueing System

Freek Verdonck$^{(\boxtimes)}$ ⓘ, Herwig Bruneel ⓘ, and Sabine Wittevrongel ⓘ

Department of Telecommunications and Information Processing (TELIN), SMACS
Research Group, Ghent University (UGent), Sint-Pietersnieuwstraat 41,
9000 Gent, Belgium
{freek.verdonck,herwig.bruneel,sabine.wittevrongel}@ugent.be

**Abstract.** In this work we look at a discrete-time multiserver queueing system where the number of available servers is distributed according to one of two geometrics. The arrival process is assumed to be general independent, the service times deterministically equal to one slot and the buffer capacity infinite. The queueing system resides in one of two states and the number of available servers follows a geometric distribution with parameter determined by the system state. At the end of a slot there is a fixed probability that the system evolves from one state to the other, with this probability depending on the current system state only, resulting in geometrically distributed sojourn times.

We obtain the probability generating function (pgf) of the system content of an arbitrary slot in steady-state, as well as the pgf of the system content at the beginning of an arbitrary slot with a given state. Furthermore we obtain an approximation of the distribution of the delay a customer experiences in the proposed queueing system. This approximation is validated by simulation and the results are illustrated with a numerical example.

**Keywords:** Queueing theory · Discrete-time · Multiserver · Geometric · System content · Delay

## 1 Introduction

In many queueing situations the number of servers is not constant. Moreover, the expected number of available servers can vary over time. In this respect, this paper focusses on a discrete-time queueing model with two system states. The number of available servers in a slot follows a geometric distribution with a parameter that is determined by the system state. The arrival process is assumed to be general and independent, service times are deterministically equal to one slot and the buffer capacity is infinite. State changes are assumed to occur according to a first-order Markov process.

Discrete-time multiserver queueing systems have received considerable attention in the past in several settings [5–8,13]. Specifically, [13] handles a multiserver queueing system with priorities. Multiserver queueing systems with batch

© Springer Nature Switzerland AG 2020
M. Gribaudo et al. (Eds.): ASMTA 2019, LNCS 12023, pp. 57–70, 2020.
https://doi.org/10.1007/978-3-030-62885-7_5

arrivals are considered in [5,6] for geometric, respectively deterministic service times, while [7,8] deal with multiserver systems with general independent arrivals and geometric and constant service times. These papers all have in common that the number of servers present in the system is constant.

Also queues with a varying number of servers have been studied in literature. In the most simple case, the number of available servers changes independently from slot to slot [1,9]. In [9] either all $m$ servers are available or none, while in [1] the number of available servers can take any value between 0 and $m$. In [4] correlation over time is introduced on the number of available servers by combining a permanently available server with an extra server with generally distributed off-times and geometrically distributed on-times. The analysis of [4], however, is limited to the system content. In the current paper, we focus on a discrete-time multiserver queueing model that is general enough to include variability and time correlation on the number of available servers, but yet simple enough to lend itself to a full queueing analysis of not only the system content but also the delay, and to have a limited number of easily interpretable model parameters. Specifically, our model considers two possible system states, each with their own geometric distributions for state sojourn time and for the number of available servers during a slot.

The main contribution of the current paper is the delay analysis for the considered queueing model. Multiserver queueing systems are notoriously hard when considering the delay analysis, especially when dealing with a varying number of available servers. Some earlier results can be found in [12] where all $m$ servers are subject to independent interruptions and no correlation is present on the server availability.

The study of this model is motivated by the many applications of queueing theory where the number of servers is not constant over time and a certain correlation exists in the number of available servers. Examples include the modelling of the airport checkin process [14] or supply chains in production facilities [11].

The outline of the paper is as follows. In the next section we provide a detailed mathematical description of the queueing model under study. In Sect. 3 we obtain the steady-state distribution of the system content at the beginning of an arbitrary slot and at the beginning of a slot with a given state. In Sect. 4 we look at the delay analysis for this queueing system. Section 5 provides a numerical example and Sect. 6 concludes the paper.

## 2  Mathematical Model

In this paper we study a discrete-time queueing system; the time horizon is divided into slots of equal length. The arrival process is general and independent and is described by

$$c(n) \triangleq \mathrm{Prob}[n \text{ customers arrive during a slot}] , \quad n \geq 0 ; \tag{1}$$

$$C(z) \triangleq \sum_{n=0}^{\infty} c(n)z^n ; \tag{2}$$

$$\lambda \triangleq \sum_{n=0}^{\infty} nc(n) = C'(1) .\tag{3}$$

If we denote by $c_k$ the number of arrivals in the $k$th slot, then the series $\{c_k\}$ is a set of independent and identically distributed (i.i.d.) stochastic variables. The service time of every customer equals 1 slot.

The system resides in state-A or state-B and the number of available servers during a slot follows a geometrical distribution, with parameter determined by the system state during that slot. Specifically, we have that

$$\mathrm{Prob}[n \text{ servers available during A-slot}] = (1 - \beta_1)\beta_1^n , \quad n \geq 0; \tag{4}$$

$$\mathrm{Prob}[n \text{ servers available during B-slot}] = (1 - \beta_2)\beta_2^n , \quad n \geq 0, \tag{5}$$

with $0 < \beta_1, \beta_2 < 1$. The expected number of available servers during an A-slot is $\frac{\beta_1}{1-\beta_1}$ and during a B-slot $\frac{\beta_2}{1-\beta_2}$. We easily obtain

$$\mathrm{Prob}[\text{more than } n \text{ servers available during A-slot}] = \beta_1^{n+1} , \quad n \geq 0; \tag{6}$$

$$\mathrm{Prob}[\text{more than } n \text{ servers available during B-slot}] = \beta_2^{n+1} , \quad n \geq 0. \tag{7}$$

If we denote by $s_{\mathrm{A},k}$ and $s_{\mathrm{B},k}$ the number of available servers in the $k$th A-slot and the $k$th B-slot respectively, then we have that the series $\{s_{\mathrm{A},k}\}$ and $\{s_{\mathrm{B},k}\}$ are two different sets of i.i.d. stochastic variables.

State changes can only occur at the end of a slot, and the probability of a state change occurring is fixed and depends solely on the current state:

$$\mathrm{Prob}[\text{A-slot is followed by B-slot}] \triangleq 1 - \alpha_1 ; \tag{8}$$

$$\mathrm{Prob}[\text{B-slot is followed by A-slot}] \triangleq 1 - \alpha_2 . \tag{9}$$

with $0 < \alpha_1, \alpha_2 < 1$. Let us introduce $\sigma$ to denote the probability that an arbitrary slot is an A-slot, then standard probability theory leads to:

$$\sigma = \frac{1 - \alpha_2}{2 - \alpha_1 - \alpha_2} .\tag{10}$$

This paper handles the steady-state situation of the queueing system as described and it is therefore assumed that the stability condition is fulfilled. For the queue to be stable it is required that the average number of arrivals during a slot is strictly smaller than the average number of customers that can be served. This can be expressed as:

$$\lambda < \frac{\sigma\beta_1}{1 - \beta_1} + \frac{(1 - \sigma)\beta_2}{1 - \beta_2} = \frac{(1 - \alpha_1)(1 - \beta_1)\beta_2 + (1 - \alpha_2)(1 - \beta_2)\beta_1}{(1 - \beta_1)(1 - \beta_2)(2 - \alpha_1 - \alpha_2)} .\tag{11}$$

We assume a First In First Out (FIFO) policy. The delay of a customer is defined as its total system time, excluding the remainder of the slot in which the customer arrives. The delay is thus always an integer number of slots and includes the service time.

The setup as described is also referred to as a Late Arrival System with Delayed Access (LAS-DA).

A schematic overview of the system can be found in Fig. 1.

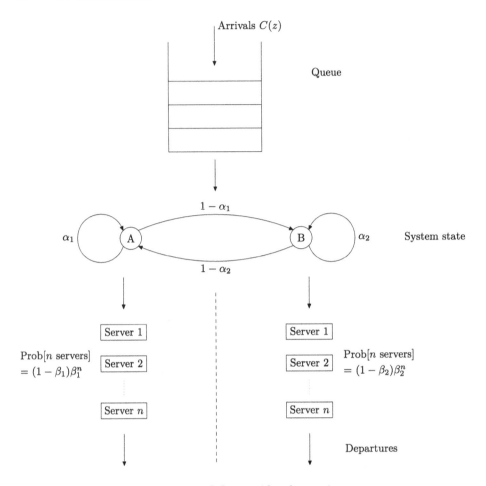

**Fig. 1.** Illustration of the considered queueing system

## 3   Analysis of System Content

In this section we first analyze the system content. Let us look at an arbitrary A-slot $S_A$ and an arbitrary B-slot $S_B$ in steady state. We denote the system content at the beginning of these slots as $u_A$ and $u_B$ respectively, while we denote the system content at the end of these arbitrary slots as $u_A^+$ and $u_B^+$ respectively. These system contents are related to each other through the following system equations:

$$u_A^+ = \max(u_A - s_A, 0) + c_A \; ; \tag{12}$$

$$u_B^+ = \max(u_B - s_B, 0) + c_B \, , \tag{13}$$

with $s_A$ and $s_B$ the number of available servers in $S_A$ and $S_B$, and with $c_A$ and $c_B$ the number of customers arriving in these respective slots. Due to the

Markovian transitions between the system states we get that in steady state there is a probability $\alpha_1$ that the slot before $S_A$ is an A-slot (and a probability $(1 - \alpha_1)$ that the slot before $S_A$ is a B-slot). If we introduce $U_A(z)$ and $U_B(z)$ as the probability generating functions (pgfs) of $u_A$ and $u_B$ respectively, we can write:

$$
U_A(z) = \alpha_1 E\left[z^{u_A^+}\right] + (1 - \alpha_1) E\left[z^{u_B^+}\right]
$$

$$
= \alpha_1 C(z) \sum_{k=0}^{\infty} \text{Prob}[u_A = k]\,\text{Prob}[s_A > k]
$$

$$
+ \alpha_1 C(z) \sum_{k=0}^{\infty} \sum_{l=0}^{k} \text{Prob}[u_A = k]\,\text{Prob}[s_A = l]\,z^{k-l}
$$

$$
+ (1 - \alpha_1) C(z) \sum_{k=0}^{\infty} \text{Prob}[u_B = k]\,\text{Prob}[s_B > k]
$$

$$
+ (1 - \alpha_1) C(z) \sum_{k=0}^{\infty} \sum_{l=0}^{k} \text{Prob}[u_B = k]\,\text{Prob}[s_B = l]\,z^{k-l}
$$

$$
= \alpha_1 C(z)\,\beta_1 U_A(\beta_1) + (1 - \alpha_1) C(z)\,\beta_2 U_B(\beta_2)
$$

$$
+ \alpha_1 C(z) \sum_{k=0}^{\infty} \text{Prob}[u_A = k]\,z^k (1 - \beta_1)\frac{1 - \left(\frac{\beta_1}{z}\right)^{k+1}}{1 - \frac{\beta_1}{z}}
$$

$$
+ (1 - \alpha_1) C(z) \sum_{k=0}^{\infty} \text{Prob}[u_B = k]\,z^k (1 - \beta_2)\frac{1 - \left(\frac{\beta_2}{z}\right)^{k+1}}{1 - \frac{\beta_2}{z}}
$$

$$
= \alpha_1 C(z)\,U_A(\beta_1)\,\beta_1 \frac{z - 1}{z - \beta_1} + (1 - \alpha_1) C(z)\,U_B(\beta_2)\,\beta_2 \frac{z - 1}{z - \beta_2}
$$

$$
+ \alpha_1 C(z)\,U_A(z)\frac{(1 - \beta_1)z}{z - \beta_1} + (1 - \alpha_1) C(z)\,U_B(z)\frac{(1 - \beta_2)z}{z - \beta_2}. \tag{14}
$$

A similar derivation can be made for the pgf $U_B(z)$ of $u_B$ leading to

$$
U_B(z) = \alpha_2 C(z)\,U_B(\beta_2)\,\beta_2 \frac{z - 1}{z - \beta_2} + (1 - \alpha_2) C(z)\,U_A(\beta_1)\,\beta_1 \frac{z - 1}{z - \beta_1}
$$

$$
+ \alpha_2 C(z)\,U_B(z)\frac{(1 - \beta_2)z}{z - \beta_2} + (1 - \alpha_2) C(z)\,U_A(z)\frac{(1 - \beta_1)z}{z - \beta_1}. \tag{15}
$$

The set of linear equations (14) and (15) can be solved for $U_A(z)$ and $U_B(z)$ which leads to the following explicit expressions:

$$
U_A(z) = \frac{(z - 1)C(z)\left\{\begin{array}{c}\beta_2(1 - \alpha_1)(z - \beta_1)U_B(\beta_2) + \beta_1\alpha_1(z - \beta_2)U_A(\beta_1) \\ + \beta_1(1 - \beta_2)(1 - \alpha_1 - \alpha_2)zC(z)\,U_A(\beta_1)\end{array}\right\}}{(z - \beta_1)(z - \beta_2) - (1 - \beta_1)(1 - \beta_2)(1 - \alpha_1 - \alpha_2)z^2 C(z)^2 - \left[\alpha_1(1 - \beta_1)(z - \beta_2) + \alpha_2(1 - \beta_2)(z - \beta_1)\right]zC(z)};
$$

$$
\tag{16}
$$

$$U_{\mathrm{B}}(z) = \frac{(z-1)C(z)\left\{\begin{array}{c}\beta_1(1-\alpha_2)(z-\beta_2)U_{\mathrm{A}}(\beta_1) + \beta_2\alpha_2(z-\beta_1)U_{\mathrm{B}}(\beta_2) \\ +\beta_2(1-\beta_1)(1-\alpha_1-\alpha_2)zC(z)\,U_{\mathrm{B}}(\beta_2)\end{array}\right\}}{\begin{array}{c}(z-\beta_1)(z-\beta_2) - (1-\beta_1)(1-\beta_2)(1-\alpha_1-\alpha_2)z^2C(z)^2 \\ -\left[\alpha_1(1-\beta_1)(z-\beta_2) + \alpha_2(1-\beta_2)(z-\beta_1)\right]zC(z)\end{array}},$$

(17)

in which two unknown constants appear, $U_{\mathrm{A}}(\beta_1)$ and $U_{\mathrm{B}}(\beta_2)$. These unknowns cannot be straightforwardly determined since the substitutions $z = \beta_1$ in (16) and $z = \beta_2$ in (17) lead to two identities. However, we can determine them by relying on the properties of pgfs, namely that they are analytical within the complex unit disk and normalized.

Let us first take a look at the denominator of $U_{\mathrm{A}}(z)$ and $U_{\mathrm{B}}(z)$. It can easily be seen that the first part, $(z-\beta_1)(z-\beta_2)$, has exactly 2 zeros within the complex unit disk. By application of Rouché's theorem we can conclude that the whole denominator also has 2 zeros within the complex unit disk (for more information on Rouché's theorem, see for example [10]). It can easily be verified that $z = 1$ is one of these zeros. Let us call the other zero $z_1$. As $U_{\mathrm{A}}(z)$ and $U_{\mathrm{B}}(z)$ are pgfs, they cannot have singularities within the complex unit disk and thus their numerators must also vanish at $z = 1$ and $z = z_1$. This is obviously the case for $z = 1$, irrespectively of the unknown constants. Expressing that the numerator must also vanish for $z = z_1$ leads to one relation linking the 2 unknowns. A second relation can be obtained by expressing the normality condition of pgfs:

$$\lim_{z \to 1} U_{\mathrm{A}}(z) = 1.$$

(18)

After applying L'Hôpital's rule and using (10), we obtain

$$\frac{\sigma\beta_1(1-\beta_2)U_{\mathrm{A}}(\beta_1) + (1-\sigma)(1-\beta_1)\beta_2 U_{\mathrm{B}}(\beta_2)}{(1-\beta_1)(1-\beta_2)\lambda + \sigma\beta_1(1-\beta_2) + (1-\sigma)(1-\beta_1)\beta_2} = 1.$$

(19)

We can thus determine the 2 unknowns in the expressions for $U_{\mathrm{A}}(z)$ and $U_{\mathrm{B}}(z)$ and obtain the pgfs of the system contents at the beginning of an arbitrary A-slot and B-slot. The system content $u$ at the beginning of an arbitrary slot is then determined by its pgf $U(z)$, with

$$U(z) = \sigma U_{\mathrm{A}}(z) + (1-\sigma)U_{\mathrm{B}}(z).$$

(20)

From the pgf of the system content we can easily obtain the (central) moments of its distribution, leading to the mean and variation of the system content. Also higher order moments can be calculated.

## 4   Delay Analysis

Now that we have expressions for the pgfs of the system contents at the beginning of an arbitrary A-slot and arbitrary B-slot, we are in a position to study the delay that a customer experiences in the queueing system. First, we condition the delay of a customer on the state of its arrival slot and on the number of customers waiting in the queue in front of it, where we exclude the customers that are receiving service at the moment of arrival. Then, we combine this with the results of the previous section to obtain an expression for the pgf of the delay of an arbitrary customer. Finally, we use the obtained pgf to derive an approximation for the tail probabilities of the delay.

## 4.1    Delay of a Customer with $k$ Customers Ahead

We look at an arbitrary customer $P_A$, arriving during an arbitrary A-slot. We introduce the stochastic variable $d_{A,k}$ for its delay, given that there are $k$ customers in front of $P_A$ in the queue at the moment of its arrival (thus excluding the customers receiving service). The corresponding pgf is $D_{A,k}(z)$. Analogously we study the arbitrary customer $P_B$, arriving during an arbitrary B-slot. Given that there are $k$ customers in the queue in front of $P_B$, its delay is denoted by the stochastic variable $d_{B,k}$, with corresponding pgf $D_{B,k}(z)$. By looking at the system state and the number of available servers during the slot after the considered customer's arrival slot, we can obtain the following relations (for $k \geq 0$):

$$D_{A,k}(z) = (1 - \alpha_1)z\,\mathrm{Prob}[s_B > k] + (1 - \alpha_1)z\sum_{l=0}^{k} D_{B,k-l}(z)\,\mathrm{Prob}[s_B = l]$$

$$+ \alpha_1 z\,\mathrm{Prob}[s_A > k] + \alpha_1 z\sum_{l=0}^{k} D_{A,k-l}(z)\,\mathrm{Prob}[s_A = l] \; ; \tag{21}$$

$$D_{B,k}(z) = (1 - \alpha_2)z\,\mathrm{Prob}[s_A > k] + (1 - \alpha_2)z\sum_{l=0}^{k} D_{A,k-l}(z)\,\mathrm{Prob}[s_A = l]$$

$$+ \alpha_2 z\,\mathrm{Prob}[s_B > k] + \alpha_2 z\sum_{l=0}^{k} D_{B,k-l}(z)\,\mathrm{Prob}[s_B = l] \; , \tag{22}$$

with $s_A$ and $s_B$ the numbers of available servers in an A-slot or B-slot. Let us now introduce some auxiliary functions:

$$D_A(x, z) \triangleq \sum_{k=0}^{\infty} D_{A,k}(z)\,x^k \; ; \tag{23}$$

$$D_B(x, z) \triangleq \sum_{k=0}^{\infty} D_{B,k}(z)\,x^k \; . \tag{24}$$

Using (21) and (22) to work out these definitions we get

$$D_A(x, z) = (1 - \alpha_1)z\sum_{k=0}^{\infty} \beta_2^{k+1}x^k + (1 - \alpha_1)z\sum_{k=0}^{\infty} x^k \sum_{l=0}^{k} D_{B,k-l}(z)\,(1 - \beta_2)\beta_2^l$$

$$+ \alpha_1 z\sum_{k=0}^{\infty} \beta_1^{k+1}x^k + \alpha_1 z\sum_{k=0}^{\infty} x^k \sum_{l=0}^{k} D_{A,k-l}(z)\,(1 - \beta_1)\beta_1^l$$

$$= \frac{(1-\alpha_1)\beta_2 z}{1-\beta_2 x} + \frac{(1-\alpha_1)(1-\beta_2)z}{1-\beta_2 x} D_B(x,z)$$
$$+ \frac{\alpha_1\beta_1 z}{1-\beta_1 x} + \frac{\alpha_1(1-\beta_1)z}{1-\beta_1 x} D_A(x,z) , \tag{25}$$

and in a similar way

$$D_B(x,z) = \frac{(1-\alpha_2)\beta_1 z}{1-\beta_1 x} + \frac{(1-\alpha_2)(1-\beta_1)z}{1-\beta_1 x} D_A(x,z)$$
$$+ \frac{\alpha_2\beta_2 z}{1-\beta_2 x} + \frac{\alpha_2(1-\beta_2)z}{1-\beta_2 x} D_B(x,z) . \tag{26}$$

The set of linear equations (25) and (26) can be solved for $D_A(x,z)$ and $D_B(x,z)$. This leads to the following explicit expressions:

$$D_A(x,z) = \frac{f_A(x,z)}{g(x,z)} ; \tag{27}$$

$$D_B(x,z) = \frac{f_B(x,z)}{g(x,z)} , \tag{28}$$

with

$$f_A(x,z) \triangleq \beta_1\beta_2 zx - [(1-\alpha_1)\beta_2 + (1-\beta_2)(1-\alpha_1-\alpha_2)\beta_1 z - \alpha_1\beta_1] z ; \tag{29}$$
$$f_B(x,z) \triangleq \beta_1\beta_2 zx - [(1-\alpha_2)\beta_1 + (1-\beta_1)(1-\alpha_1-\alpha_2)\beta_2 z - \alpha_2\beta_2] z , \tag{30}$$

and

$$g(x,z) \triangleq -1 + [\alpha_1(1-\beta_1) + \alpha_2(1-\beta_2)] z + (1-\beta_1)(1-\beta_2)(1-\alpha_1-\alpha_2)z^2$$
$$+ [(1-\alpha_1 z)\beta_2 + (1-\alpha_2 z)\beta_1 + (\alpha_1+\alpha_2)\beta_1\beta_2 z] x - \beta_1\beta_2 x^2. \tag{31}$$

We can consider $D_A(x,z)$ and $D_B(x,z)$ as rational functions in $x$ with numerator of degree 1 and denominator of degree 2. A partial fraction expansion can be made based on the poles in $x$ which we denote as $x_1$ and $x_2$ and assume to be distinct. We can then rewrite (27) and (28) as

$$D_A(x,z) = \sum_{i=1}^{2} \frac{f_A(x_i,z)}{g_x(x_i,z)(x-x_i)} ; \tag{32}$$

$$D_B(x,z) = \sum_{i=1}^{2} \frac{f_B(x_i,z)}{g_x(x_i,z)(x-x_i)} , \tag{33}$$

with

$$g_x(x,z) \triangleq \frac{\partial}{\partial x} g(x,z) . \tag{34}$$

We obtain an expression for $D_{A,k}(z)$ by evaluating the $k$th derivative with respect to $x$ of $D_A(x,z)$ at $x=0$:

$$D_{A,k}(z) = \frac{1}{k!} \frac{\partial^k}{\partial x^k} D_A(x,z)\Big|_{x=0}$$

$$= \sum_{i=1}^{2} \frac{-f_A(x_i, z)}{g_x(x_i, z) \, x_i^{k+1}} . \tag{35}$$

In a similar way we find the following expression for $D_{B,k}(z)$:

$$D_{B,k}(z) = \sum_{i=1}^{2} \frac{-f_B(x_i, z)}{g_x(x_i, z) \, x_i^{k+1}} . \tag{36}$$

## 4.2   Delay of an Arbitrary Customer

We consider the arbitrary packet $P_A$, arriving in the system during the A-slot $S_A$ and we denote the number of customers in the queue at its moment of arrival by the stochastic variable $t_A$, with corresponding pgf $T_A(z)$. Upon arrival, the customers waiting in the queue are those that were present in the queueing system at the beginning of $S_A$, minus those that receive service during $S_A$ and plus those that arrived during $S_A$, but before the arrival of $P_A$. The pgf $F(z)$ of this last number of arrivals is well known in the literature, see e.g. [2]:

$$F(z) = \frac{C(z) - 1}{\lambda(z - 1)} . \tag{37}$$

We get for $T_A(z)$:

$$T_A(z) = F(z) \left\{ \sum_{k=0}^{\infty} \text{Prob}[u_A = k] \, \text{Prob}[s_A > k] \right.$$

$$\left. + \sum_{k=0}^{\infty} \text{Prob}[u_A = k] \sum_{l=0}^{k} \text{Prob}[s_A = l] \, z^{k-l} \right\}$$

$$= F(z) \left\{ \sum_{k=0}^{\infty} \text{Prob}[u_A = k] \, \beta_1^{k+1} \right.$$

$$\left. + \sum_{k=0}^{\infty} \text{Prob}[u_A = k] \, z^k \sum_{l=0}^{k} (1 - \beta_1) \left( \frac{\beta_1}{z} \right)^l \right\}$$

$$= F(z) \left\{ \beta_1 U_A(\beta_1) + (1 - \beta_1) \sum_{k=0}^{\infty} \text{Prob}[u_A = k] \, \frac{z^{k+1} - \beta_1^{k+1}}{z - \beta_1} \right\}$$

$$= F(z) \frac{(1 - \beta_1) z U_A(z) + (z - 1) \beta_1 U_A(\beta_1)}{z - \beta_1} . \tag{38}$$

In a similar manner we can define $t_B$ as the queue content seen by an arbitrary customer $P_B$, arriving during a B-slot. The corresponding pgf $T_B(z)$ is given by

$$T_B(z) = F(z) \frac{(1 - \beta_2) z U_B(z) + (z - 1) \beta_2 U_B(\beta_2)}{z - \beta_2} . \tag{39}$$

The arrival slot of an arbitrary packet $P$ is an A-slot with probability $\sigma$ and a B-slot with probability $(1 - \sigma)$. We can therefore express the pgf of its delay $W(z)$ as

$$W(z) = \sigma \sum_{k=0}^{\infty} \text{Prob}[t_A = k] \, D_{A,k}(z) + (1 - \sigma) \sum_{k=0}^{\infty} \text{Prob}[t_B = k] \, D_{B,k}(z) . \quad (40)$$

Substitution of (35) and (36) into the above expression yields

$$W(z) = \sigma \sum_{i=1}^{2} \frac{-f_A(x_i, z)}{g_x(x_i, z) \, x_i} T_A\left(\frac{1}{x_i}\right) + (1 - \sigma) \sum_{i=1}^{2} \frac{-f_B(x_i, z)}{g_x(x_i, z) \, x_i} T_B\left(\frac{1}{x_i}\right) . \quad (41)$$

The above equation is fully determined, the $x_i$ are functions of $z$, but can be easily obtained as they are the roots of a quadratic equation. However, it is not easy to invert this pgf. In the following subsection we use these results to obtain a tail approximation of the delay of an arbitrary customer.

### 4.3   Tail Approximation

We can use the technique of the dominant singularity, see e.g. [3,15], to obtain the tail distribution of the delay of an arbitrary customer. For sufficiently large $k$ we have that

$$\text{Prob}[\text{delay} = k \text{ slots}] \approx -\frac{w_0}{z_0} z_0^{-k} ; \quad (42)$$

$$\text{Prob}[\text{delay} > k \text{ slots}] \approx -\frac{w_0}{z_0(z_0 - 1)} z_0^{-k} , \quad (43)$$

with $z_0$ the pole of $W(z)$ with the smallest modulus and with

$$w_0 = \lim_{z \to z_0} [W(z)(z - z_0)] . \quad (44)$$

Note that $z_0$ is real-valued and larger than 1 and that $w_0$ is real-valued and negative. As the $x_i$ are non-zero and distinct we have that $z_0$ can only be found as a pole of $T_A\left(\frac{1}{x_i}\right)$, or equivalently as a pole of $T_B\left(\frac{1}{x_i}\right)$. In view of (16), (37) and (38), $z_0$ must be found as a pole of $C\left(\frac{1}{x_i}\right)$ or as a zero of

$$f(z) = (1 - \beta_1 x_i)(1 - \beta_2 x_i) - (1 - \beta_1)(1 - \beta_2)(1 - \alpha_1 - \alpha_2)C\left(\frac{1}{x_i}\right)^2$$

$$- [\alpha_1(1 - \beta_1)(1 - \beta_2 x_i) + \alpha_2(1 - \beta_2)(1 - \beta_1 x_i)] C\left(\frac{1}{x_i}\right) . \quad (45)$$

## 5   Numerical Examples

In this section we illustrate the method developed in this paper with some numerical examples. For the arrivals we take a Poisson process:

$$C(z) = e^{\lambda(z-1)} . \quad (46)$$

In an initial example we take the following values for the parameters: $\alpha_1 = 0.6$, $\alpha_2 = 0.7$, $\beta_1 = 0.4$ and $\beta_2 = 0.6$. The average number of servers available in an arbitrary slot equals 1.14 and thus the system is stable if $\lambda < 1.14$. In Fig. 2 we plot the average system content in function of the arrival intensity $\lambda$ for this system based on the method developed in this paper, as well as based on simulation. It must be noted that for the simulation the required computation time is much larger than the time required to compute the unknowns $U_A(\beta_1)$ and $U_B(\beta_2)$, which involves finding the root of a non-polynomial function within the complex unit disk and solving a set of 2 linear equations. The system content shows an expected vertical asymptote for $\lambda \to 1.14$. The simulation validates the results obtained by the method of this paper.

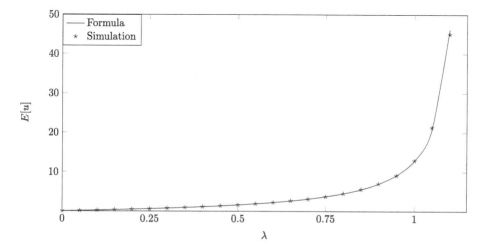

**Fig. 2.** Average system content in function of $\lambda$, for $\alpha_1 = 0.6$, $\alpha_2 = 0.7$, $\beta_1 = 0.4$ and $\beta_2 = 0.6$.

In Fig. 3 we look at the delay characteristics for this example. We set $\lambda = 1$ and we plot the probability that the delay of a customer equals $k$ slots for increasing $k$. The mean system content in this situation equals 12.88 customers. We see that already for small $k$ the obtained tail approximation is very close to the simulation results. For large $k$ the simulation would need to run for a very long time in order to get reliable results, while our formula immediately gives an excellent approximation for all $k$.

In a second example we introduce a larger difference between the two states: $\alpha_1 = 0.61$, $\alpha_2 = 0.3$, $\beta_1 = 0.1$ and $\beta_2 = 0.75$. The parameters have been chosen in such a way that the expected number of servers available during an arbitrary slot is the same as in the previous example. In the A-state there is now a high probability that no servers are available, while in state-B we expect on average 3 available servers. However, the system does not reside for long periods in the B-state as $\alpha_2$ is small. In Fig. 4 the system content is plotted in function of the

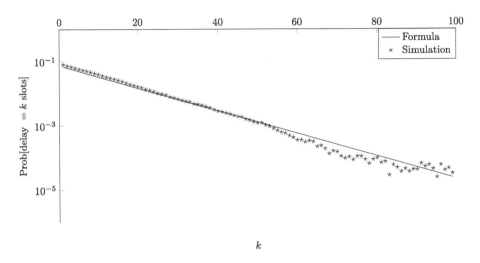

**Fig. 3.** Delay characteristics for $\lambda = 1$, $\alpha_1 = 0.6$, $\alpha_2 = 0.7$, $\beta_1 = 0.4$ and $\beta_2 = 0.6$.

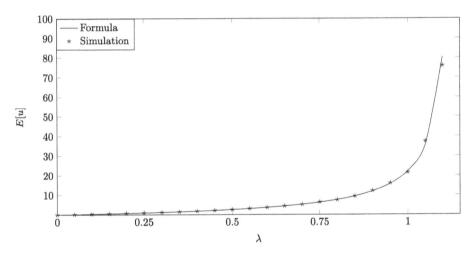

**Fig. 4.** Average system content in function of $\lambda$, for $\alpha_1 = 0.61$, $\alpha_2 = 0.3$, $\beta_1 = 0.1$ and $\beta_2 = 0.75$.

arrival intensity $\lambda$. Note that the vertical axis now goes until 100. The curve of the system content shows the same shape and has the same vertical asymptote for $\lambda \to 1.14$, but for a given $\lambda$ the system content is higher as compared to the first situation. This is expected in view of the increased irregularity in the service process for the second example. The difference is also larger for higher $\lambda$.

We now also look at the delay characteristics for this situation. We choose $\lambda = 0.9157$ in order to have the same mean system content of 12.88 as in the previous example. The delay characteristics are plotted in Fig. 5. The slope of the delay curve in this situation is less steep than before, the characteristic

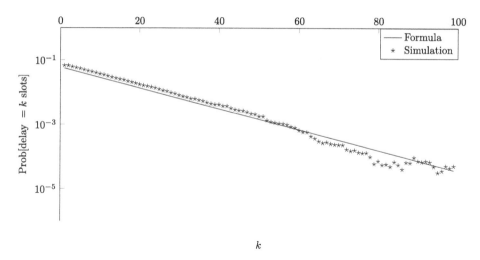

**Fig. 5.** Delay characteristics for $\lambda = 0.9157$, $\alpha_1 = 0.61$, $\alpha_2 = 0.3$, $\beta_1 = 0.1$ and $\beta_2 = 0.75$.

singularity being closer to 1. In particular we now have $z_0 = 1.0769$, while for Fig. 3 we had $z_0 = 1.0843$. This means that even though the mean system content remains the same, the delay characteristics are different. For large values of $k$, there is a higher probability that the delay of a customer exceeds $k$ as compared to the situation of Fig. 3. This is again in accordance with the increased irregularity in the service process.

## 6    Conclusion

In this paper we have studied an all geometric discrete-time multiserver queueing system. The system resides in one of two different states, and does so for a geometrically distributed number of slots, with a different parameter for each state. The number of servers available during a slot also follows a geometric distribution, with a parameter depending on the system state. The model can be used in many applications of queueing theory where the expected number of available servers fluctuates over time and is fairly simple in terms of the number of parameters that needs to be matched. We have obtained expressions for the probability generating functions of the system content at the beginning of an arbitrary slot and at the beginning of an arbitrary slot with a given state.

Furthermore we have obtained an approximation for the tail probabilities of the delay that an arbitrary customer experiences in the queueing system. This approximation is based on the theory of the dominant singularity. Numerical examples have shown that our tail approximation is also accurate for smaller delay values. The numerical examples have further illustrated that more variation in the number of available servers leads to higher system contents. Moreover, for a given system content, there is a higher probability that a customer experiences larger delays when more variability is present in the system.

# References

1. Bruneel, H.: A general model for the behavior of infinite buffers with periodic service opportunities. Eur. J. Oper. Res. **16**(1), 98–106 (1984)
2. Bruneel, H., Kim, B.G.: Discrete-Time Models for Communication Systems Including ATM. Kluwer Academic Publishers Group (1993)
3. Bruneel, H., Steyaert, B., Desmet, E., Petit, G.: Analytic derivation of tail probabilities for queue lengths and waiting times in ATM multiserver queues. Eur. J. Oper. Res. **76**(3), 563–572 (1994)
4. Bruneel, H., Wittevrongel, S.: Analysis of a discrete-time single-server queue with an occasional extra server. Perform. Eval. **116**, 119–142 (2017)
5. Chaudhry, M.L., Gupta, U., Goswami, V.: Modeling and analysis of discrete-time multiserver queues with batch arrivals: $GI^X$/Geom/m. INFORMS J. Comput **13**(3), 172–180 (2001)
6. Chaudhry, M.L., Kim, N.: A complete and simple solution for a discrete-time multiserver queue with bulk arrivals and deterministic service times. Oper. Res. Lett. **31**(2), 101–107 (2003)
7. Gao, P., Wittevrongel, S., Bruneel, H.: Discrete-time multiserver queues with geometric service times. Comput. Oper. Res. **31**(1), 81–99 (2004)
8. Gao, P., Wittevrongel, S., Walraevens, J., Moeneclaey, M., Bruneel, H.: Calculation of delay characteristics for multiserver queues with constant service times. Eur. J. Oper. Res. **199**(1), 170–175 (2009)
9. Georganas, N.D.: Buffer behavior with Poisson arrivals and bulk geometric service. IEEE Trans. Commun. **24**, 938–940 (1976)
10. Gonzalez, M.: Classical Complex Analysis. CRC Press (1991)
11. Kerbache, L., Smith, J.: Queueing networks and the topological design of supply chain systems. Int. J. Prod. Econ. **91**, 251–272 (2004)
12. Laevens, K., Bruneel, H.: Delay analysis for discrete-time queueing systems with multiple randomly interrupted servers. Eur. J. Oper. Res. **85**, 161–177 (1995)
13. Laevens, K., Bruneel, H.: Discrete-time multiserver queues with priorities. Perform. Eval. **33**, 249–275 (1998)
14. Stolletz, R.: Analysis of passenger queues at airport terminals. Res. Transp. Bus. Manag. **1**(1), 144–149 (2011)
15. Woodside, C., Ho, E.: Engineering calculation of overflow probabilities in buffers with Markov-interrupted service. IEEE Trans. Commun. **35**(12), 1272–1277 (1987)

# Dealing with Dependence in Stochastic Network Calculus – Using Independence as a Bound

Paul Nikolaus[1(✉)], Jens Schmitt[1], and Florin Ciucu[2]

[1] Distributed Computer Systems (DISCO) Lab, TU Kaiserslautern, Kaiserslautern, Germany
{nikolaus,jschmitt}@cs.uni-kl.de
[2] Computer Science Department, University of Warwick, Coventry, UK
f.ciucu@warwick.ac.uk

**Abstract.** Computing probabilistic end-to-end delay bounds is an old, yet still challenging problem. Stochastic network calculus enables closed-form delay bounds for a large class of arrival processes. However, it encounters difficulties in dealing with dependent flows, as standard techniques require to apply Hölder's inequality. In this paper, we present an alternative bounding technique that, under specific conditions, treats them as if flows were independent. We show in two case studies that it often provides better delay bounds while simultaneously significantly improving the computation time.

## 1 Introduction

Stochastic network calculus (SNC) is a versatile framework to compute stochastic per-flow delay bounds. Developed as a deterministic worst-case analysis in the 1990s by Cruz [6,7], stochastic extensions of network calculus emerged quickly thereafter. It allows for closed-form solutions for a broad class of arrival and service processes. In [18], it has been shown that the SNC branch using moment-generating functions [4,11] provides tighter bounds than the approach using envelope functions [5,8,12], as it leverages the independence of arrival flows. However, many results limit the end-to-end analysis to pure tandem topologies.

Analyzing more general networks requires to consider also dependent flows at some points in the network, as the sharing of a resource clearly has a mutual impact on the flows' output behavior. Therefore, if we want to obtain the moment-generating function (MGF) of aggregated, yet dependent arrival processes $A_1(s,t)$ and $A_2(s,t)$, we typically invoke Hölder's inequality:

$$\mathrm{E}\left[e^{\theta(A_1(s,t)+A_2(s,t))}\right] \leq \mathrm{E}\left[e^{p\theta A_1(s,t)}\right]^{1/p} \cdot \mathrm{E}\left[e^{q\theta A_2(s,t)}\right]^{1/q},$$

where $0 \leq s \leq t$, $\theta > 0$, and $\frac{1}{p} + \frac{1}{q} = 1$, $p,q \in [1,\infty]$. Hölder's inequality is completely oblivious of the actual dependence structure, thus it often leads to

© Springer Nature Switzerland AG 2020
M. Gribaudo et al. (Eds.): ASMTA 2019, LNCS 12023, pp. 71–84, 2020.
https://doi.org/10.1007/978-3-030-62885-7_6

very conservative bounds. Furthermore, it places the burden of an additional, nonlinear parameter for each application to optimize.

Dependence of arrivals does not have to be a negative property per se. Taking advantage of the information about the dependence structure to improve upon the bounds has been attempted before. In [9], the functional dependence is estimated using a copula-based approach. In our work, we investigate a simpler alternative, using the independent scenario as an upper bound. To that end, we rely on a characteristic called negative dependence. We explain the main idea with the help of the following, simplistic example.

Consider a single time slot assuming two arrival processes, $A_1$ and $A_2$, that are multiplexed at one server. Both arrivals send one packet, each independently with probability $p \in (0,1)$, and the server serves one packet but strictly prioritizes $A_2$. Clearly, their two outputs, $D_1$ and $D_2$, are strongly dependent, as an arrival of the prioritized flow forces the other one to wait in the queue. Simply put, if one flow get a larger share of the server's capacity, the other is more likely to have less output. For the joint distribution of the output, we have by assumption for the departures both being equal to 1, that $P(D_1 = 1, D_2 = 1) = 0$. On the other hand, we compute for the product distribution by a simple conditioning, that $P(D_1 = 1) \cdot P(D_2 = 1) = (p \cdot (1-p)) \cdot (1-p) > 0$. Hence, if we deliberately forego the knowledge about the dependence structure, we only obtain an upper bound, yet, it allows us to consider just the marginal distributions.

The rest of the paper is structured as follows. Section 2 introduces the necessary network calculus definitions and notations as well as some preliminary results. Section 3 contains the main results obtained in two case studies assuming a conjecture on dependence. The numerical evaluation is presented in Sect. 4. Section 5 discusses the paper.

## 2    Definitions and Modeling Assumptions

### 2.1    SNC Background and Notation

We use the MGF-based SNC in order to bound the probability that the delay exceeds a given value $T$. The MGF bound on a probability is established by applying Chernoff's bound [16]

$$P(X > a) \le e^{-\theta a} \mathrm{E}[e^{\theta X}], \quad \theta > 0.$$

We define an *arrival flow* by the stochastic process $A$ with discrete time space $\mathbb{N}$ and continuous state space $\mathbb{R}_0^+$ as $A(s,t) := \sum_{i=s+1}^{t} a_i$, with $a_i$ as the traffic increment process in time slot $i$. Network calculus provides an elegant system-theoretic analysis by employing min-plus algebra.

**Definition 1 (Convolution and Deconvolution in Min-Plus Algebra [2]).** *Let $x(s,t)$ and $y(s,t)$ be real-valued, bivariate functions. The* min-plus convolution *of $x$ and $y$ is defined as*

$$x \otimes y\,(s,t) := \inf_{s \le i \le t} \{x(s,i) + y(i,t)\}.$$

*The* min-plus deconvolution *of x and y is defined as*

$$x \oslash y\,(s,t) := \sup_{0 \leq i \leq s} \{x(i,t) - y(i,s)\}.$$

The characteristics of the service process are captured by the notion of a dynamic $S$-server.

**Definition 2 (Dynamic $S$-Server** [4]**).** *Assume a service element has an arrival flow $A$ as its input and the respective output is denoted by $D$. Let $S(s,t)$, $0 \leq s \leq t$, be a stochastic process that is nonnegative and increasing in $t$. The service element is a* dynamic $S$-server *iff for all $t \geq 0$ it holds that*

$$D(0,t) \geq A \otimes S\,(0,t) = \inf_{0 \leq s \leq t} \{A(0,s) + S(s,t)\}.$$

The analysis is based on a per-flow perspective. That is, we consider a certain flow, the so-called *flow of interest* (foi). Throughout this paper, for the sake of simplicity, we assume the servers' scheduling to be arbitrary multiplexing [19]. That is, if flow $f_2$ is prioritized over flow $f_1$, the leftover service at a dynamic $S$-server for the corresponding arrival $A_1$ is $S_{\text{l.o.}}(s,t) = [S(s,t) - A_2(s,t)]^+$. Therefore, we require the server to be work-conserving.

**Definition 3 (Work-Conserving Server** [4,11]**).** *For any $t \geq 0$ let $\tau := \sup \{s \in [0,t] : D(0,s) = A(0,s)\}$ be the beginning of the last backlogged period before $t$. Assume again the service $S(s,t)$, $0 \leq s \leq t$, to be a stochastic process that is nonnegative and increasing in $t$ with $S(\tau,\tau) = 0$. A server is said to be* work-conserving *if for any fixed sample path the server is not idle and uses the entire available service, i.e., $D(0,t) = D(0,\tau) + S(\tau,t)$.*

**Definition 4 (Virtual Delay).** *The* virtual delay *at time $t \geq 0$ is defined as*

$$d(t) := \inf \{\tau \geq 0 : A(0,t) \leq D(0,t+\tau)\}.$$

It can briefly be described as the time it takes for the cumulated departures to "catch up with" the cumulated arrivals.

**Theorem 1 (Output and Delay Bound)** [4,11]. *Consider an arrival process $A(s,t)$ with dynamic $S$-server $S(s,t)$.*
*The departure process $D$ is upper bounded for any $0 \leq s \leq t$ according to*

$$D(s,t) \leq A \oslash S\,(s,t). \tag{1}$$

*The delay at $t \geq 0$ is upper bounded by*

$$d(t) \leq \inf \{\tau \geq 0 : A \oslash S\,(t+\tau,t) \leq 0\}.$$

We focus on the analogue of Theorem 1 for moment-generating functions:

**Theorem 2 (Output and Delay MGF-Bound** [3,11]**).** *For the assumptions as in Theorem 1, we obtain:*

*The MGF of the departure process D is upper bounded for any $0 \leq s \leq t$ according to*

$$E\left[e^{\theta D(s,t)}\right] \leq E\left[e^{\theta(A \oslash S\,(s,t))}\right].$$

*The violation probability of a given stochastic delay bound $T \geq 0$ at time $t \geq 0$ is bounded by*

$$P(d(t) > T) \leq E\left[e^{\theta(A \oslash S\,(t+T,t))}\right]. \tag{2}$$

In the following definition, we introduce $(\sigma, \rho)$-constraints [4] as they enable us to give time-independent, stationary bounds under stability.

**Definition 5 ($(\sigma, \rho)$-Bound [4]).** *An arrival flow is $(\sigma_A, \rho_A)$-bounded for some $\theta > 0$, if for all $0 \leq s \leq t$*

$$E\left[e^{\theta A(s,t)}\right] \leq e^{\theta(\rho_A(\theta)(t-s) + \sigma_A(\theta))}.$$

## 2.2   Negative Dependence and Acceptable Random Variables

As we discussed in the introduction, we would like to bound the joint distribution of two random variables by their respective product distribution. This concept was captured in the 1960s by Lehmann and his notion of negative dependence.

**Definition 6 (Negative Dependence [14]).** *A finite family of random variables $\{X_1, \ldots, X_n\}$ is said to be* negatively (orthant) dependent (ND) *if the two following inequalities hold:*

$$P(X_1 \leq x_1, \ldots, X_n \leq x_n) \leq \prod_{i=1}^{n} P(X_i \leq x_i),$$

$$P(X_1 > x_1, \ldots, X_n > x_n) \leq \prod_{i=1}^{n} P(X_i > x_i),$$

*for all real numbers $x_1, \ldots, x_n$.*

The following lemma shows how this characteristic can be used directly in the context of MGFs.

**Lemma 1 ([13, 20]).** *If $\{X_1, \ldots, X_n\}$ is a set of ND random variables, then for any $\theta > 0$,*

$$E\left[e^{\theta \sum_{i=1}^{n} X_i}\right] \leq \prod_{i=1}^{n} E[e^{\theta X_i}]. \tag{3}$$

In other words, treating the aggregate of ND random variables as if they were independent yields an upper bound for the respective MGFs. Random variables that suffice Eq. (3) are called "acceptable" [1], but are studied in an unrelated context.

Proving that random variables are negatively dependent is a challenging task. Some results exist, e.g., the multinomial and multivariate hypergeometric distribution are ND, or the "Zero-One Lemma" [10], which proves the property for $X_1, \ldots, X_n \in \{0, 1\}$ such that $\sum_i X_i = 1$. This means that the output processes in the example in Sect. 1 are indeed ND. Furthermore, it has been shown a related result in [13] that a permutation distribution, and therefore random sampling without replacement, is ND. In our context, this provides a result for a single time slot. In the following, we confine ourselves to conjecture this property for intervals.

*Conjecture 1.* Let two independent flows with according arrival processes $A_1$ and $A_2$ traverse a work-conserving server with finite capacity. Further, both arrival processes have iid increments.

Then, we assume their respective output processes $D_1(s, t)$ and $D_2(s, t)$ to be ND for all $0 \leq s \leq t$.

We do not have a proof but Conjecture 1 held in all our experiments using $10^6$ samples to estimate the joint and product (C)CDFs, respectively: For two flows at one server, we tried over 5500 different combinations of intervals, $x_1$, $x_2$, (as in the CDF), utilizations (between 0.4 and 0.9), and random packet sizes that were drawn from either exponential, Weibull, Gumbel, or log-normal distribution.

The focus on the same interval for both process is important, as the following, admittedly simplifying, argument suggests: Assume the high priority (HP) flow to send a lot of packets consecutively, i.e., the low priority (LP) flow has no output in this period and queues all its packets. Then, it is more likely for the LP flow to have outputs when the HP flow stops sending, as it is more likely for it to have queued packets.

## 3   Independence as a Bound

In this section, we investigate two case studies to show in which part of the analysis we exploit the negative dependence.

In the following, we consider the flow $f_1$ to be the flow of interest (foi) whose delay we stochastic upper bound. All arrival processes $A_i$ are assumed to be discrete time and to have iid increments and all servers $S_j$ are work-conserving and provide a constant rate $c_j \geq 0$. To simplify notation, we denote by $D_i^{(j)}$ the output of flow $i$ at server $j$.

### 3.1   Diamond Network

In this case study, we consider the topology in Fig. 1. Assume the foi to have the lowest priority and $f_3$ to have the highest priority. By SNC literature [5, 11], the service provided for the flow of interest, also known as the network service curve, can be described by

$$S_{\text{net}} = \left[ S_1 - \left( \left( \left( A_2 \oslash [S_4 - A_3]^+ \right) \oslash S_2 \right) + \left( (A_3 \oslash S_4) \oslash S_3 \right) \right) \right]^+ .$$

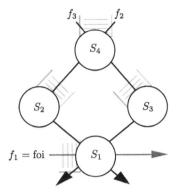

**Fig. 1.** Diamond network.

Since Conjecture 1 is made on output processes, we postpone the application of the output bound in Eq. (1) by keeping the exact output at first. That is, we start with

$$S_{\text{net}} = \left[ S_1 - \left( D_2^{(2)} + D_3^{(3)} \right) \right]^+, \tag{4}$$

use then the conjecture to bound the MGF of the aggregate by their product (Eq. (3)), and apply the output bound in a final step.

The probability that the delay process $d(t)$ exceeds a value $T \geq 0$ is upper bounded by

$$P(d(t) > T)$$

$$\overset{(2)}{\leq} E\left[ e^{\theta A_1 \oslash S_{\text{net}}(t+T,t)} \right]$$

$$\leq \sum_{\tau_1=0}^{t} E\left[ e^{\theta(A_1(\tau_1,t) - S_{\text{net}}(\tau_1,t+T))} \right]$$

$$\overset{(4)}{=} \sum_{\tau_1=0}^{t} E\left[ e^{\theta A_1(\tau_1,t)} \right] E\left[ e^{-\theta\left[ S_1 - \left( D_2^{(2)} + D_3^{(3)} \right) \right]^+ (\tau_1,t+T)} \right]$$

$$\leq \sum_{\tau_1=0}^{t} E\left[ e^{\theta A_1(\tau_1,t)} \right] e^{-\theta c_1(t+T-\tau_1)} E\left[ e^{\theta\left( D_2^{(2)} + D_3^{(3)} \right)(\tau_1,t+T)} \right], \tag{5}$$

where we used Theorem 2 in the first inequality and the Union bound in the line below. Since the flows $f_2$ and $f_3$ share the server $S_4$, their according output processes $D_2^{(4)}$ and $D_3^{(4)}$ are dependent and, as a consequence, $D_2^{(2)}$ and $D_3^{(3)}$, as well. However, by the conjecture above, we assume that the resource sharing at $S_4$ indicates that the dependence on $[\tau_1, t+T]$ is negative which, in turn, is the reason why we upper bound their joint MGF by the product of the marginal MGFs.

This can be interpreted as if we analyzed a new system, where the server $S_4$ would be split into two servers. That is, one provides the same service as the

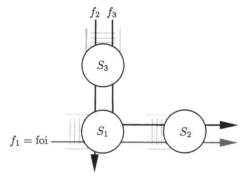

**Fig. 2.** The $\mathbb{L}$.

original (for the high priority flow $f_3$), and the other provides the leftover service $[S_4' - A_3']^+$, where $S_4'$ has the same service rate as $S_4$ and $A_3'$ is a new arrival process, but with the same distribution as $A_3$.

Hence, the second factor is upper bounded by

$$
\begin{aligned}
&\mathrm{E}\left[e^{\theta\left(D_2^{(2)}+D_3^{(3)}\right)(\tau_1,t+T)}\right] \\
&\leq \mathrm{E}\left[e^{\theta D_2^{(2)}(\tau_1,t+T)}\right]\mathrm{E}\left[e^{\theta D_3^{(3)}(\tau_1,t+T)}\right] \\
&\leq \mathrm{E}\left[e^{\theta\left(\left(A_2\oslash[S_4-A_3]^+\right)\oslash S_2\right)(\tau_1,t+T)}\right]\mathrm{E}\left[e^{\theta\left((A_3\oslash S_4)\oslash S_3\right)(\tau_1,t+T)}\right].
\end{aligned}
$$

Further assuming all $A_i$ to be $(\sigma_A,\rho_A)$-bounded yields a closed-form for the delay bound under stability:

$$
\begin{aligned}
\mathrm{P}(d(t)>T)\leq{}&\frac{e^{\theta\left(\left(\rho_{A_2}(\theta)+\rho_{A_3}(\theta)-c_1\right)T+\sigma_1(\theta)+\sigma_{A_2}(\theta)+2\sigma_{A_3}(\theta)\right)}}{1-e^{\theta\left(\rho_{A_1}(\theta)+\rho_{A_2}(\theta)+\rho_{A_3}(\theta)-c_1\right)}} \\
&\cdot\frac{1}{1-e^{\theta\left(\rho_{A_2}(\theta)-c_2\right)}}\cdot\frac{1}{1-e^{\theta\left(\rho_{A_3}(\theta)-c_3\right)}} \\
&\cdot\frac{1}{1-e^{\theta\left(\rho_{A_2}(\theta)+\rho_{A_3}(\theta)-c_4\right)}}\cdot\frac{1}{1-e^{\theta\left(\rho_{A_3}(\theta)-c_4\right)}}.
\end{aligned}
$$

For detailed calculations we refer to Appendix A.1.

In contrast, standard techniques proceed at Eq. (5) by applying the output Bound Eq. (1) immediately and continue with Hölder's inequality to deal with the dependence.

## 3.2    The $\mathbb{L}$

In this case study, we analyze the topology in Fig. 2. The foi has the lowest priority and $f_2$ the highest. Similarly to Subsect. 3.1, we assume the outputs processes of $f_2$ and $f_3$ to be ND, based on Conjecture 1. Here, the network service curve is

$$S_{\text{net}} = \left[ \left( [S_1 - (A_2 \oslash S_3)]^+ \otimes S_2 \right) - \left( A_3 \oslash [S_3 - A_2]^+ \right) \right]^+ .$$

Again, we postpone the output bounding and start with

$$S_{\text{net}} = \left[ \left( \left[ S_1 - D_2^{(3)} \right]^+ \otimes S_2 \right) - D_3^{(3)} \right]^+ . \tag{6}$$

The crucial difference is that, in order to obtain the delay bound for the foi, the so-called min-plus convolution has to be applied to the service processes of $S_1$ and $S_2$ forcing us to analyze the output processes at different intervals:

$$P(d(t) > T)$$

$$\overset{(2)}{\leq} \mathrm{E}\left[ e^{\theta A_1 \oslash S_{\text{net}}(t+T,t)} \right]$$

$$\leq \sum_{\tau_1=0}^{t} \mathrm{E}\left[ e^{\theta(A_1(\tau_1,t)-S_{\text{net}}(\tau_1,t+T))} \right]$$

$$\overset{(6)}{=} \sum_{\tau_1=0}^{t} \mathrm{E}\left[ e^{\theta A_1(\tau_1,t)} \right] \mathrm{E}\left[ e^{-\theta \left[ \left( \left[ S_1 - D_2^{(3)} \right]^+ \otimes S_2 \right) - D_3^{(3)} \right]^+ (\tau_1,t+T)} \right]$$

$$\leq \sum_{\tau_1=0}^{t} \mathrm{E}\left[ e^{\theta A_1(\tau_1,t)} \right] \mathrm{E}\left[ e^{\theta D_3^{(3)}(\tau_1,t+T)} e^{-\theta \left( \left[ S_1 - D_2^{(3)} \right]^+ \otimes S_2 \right)(\tau_1,t+T)} \right]$$

$$\leq \sum_{\tau_1=0}^{t} \mathrm{E}\left[ e^{\theta A_1(\tau_1,t)} \right] \sum_{\tau_2=\tau_1}^{t+T} \mathrm{E}\left[ e^{\theta D_3^{(3)}(\tau_1,t+T)} e^{-\theta \left[ S_1 - D_2^{(3)} \right]^+ (\tau_1,\tau_2)} e^{-\theta S_2(\tau_2,t+T)} \right]$$

$$\leq \sum_{\tau_1=0}^{t} \mathrm{E}\left[ e^{\theta A_1(\tau_1,t)} \right] \sum_{\tau_2=\tau_1}^{t+T} e^{-\theta c_1 \cdot (\tau_2-\tau_1)} e^{-\theta c_2 \cdot (t+T-\tau_2)} \mathrm{E}\left[ e^{\theta D_3^{(3)}(\tau_1,t+T)} e^{\theta D_2^{(3)}(\tau_1,\tau_2)} \right],$$

where we used the Union bound for each application of the convolution/deconvolution. This scenario is not covered by Conjecture 1 (see also the discussion at the end of Subsect. 2.2). Our work-around is to leverage the monotonicity of $D_2^{(3)}$:

$$\mathrm{E}\left[ e^{\theta D_3^{(3)}(\tau_1,t+T)} e^{\theta D_2^{(3)}(\tau_1,\tau_2)} \right] \leq \mathrm{E}\left[ e^{\theta D_3^{(3)}(\tau_1,t+T)} e^{\theta D_2^{(3)}(\tau_1,t+T)} \right].$$

The rest of the analysis employs similar techniques as for the diamond network. See also Appendix A.2. Under the assumption of $(\sigma_A, \rho_A)$-bounded arrivals, we again obtain a closed form for a bound on the delay's violation probability under stability:

$$P(d(t) > T) \leq \frac{e^{\theta\left( (\rho_{A_2}(\theta) + \rho_{A_3}(\theta) - \min\{c_1,c_2\}) \cdot T + \sigma_{A_1}(\theta) + 2\sigma_{A_2}(\theta) + \sigma_{A_3}(\theta) \right)}}{1 - e^{\theta\left( \rho_{A_1}(\theta) + \rho_{A_2}(\theta) + \rho_{A_3}(\theta) - \min\{c_1,c_2\} \right)}}$$

$$\cdot \frac{1}{1 - e^{\theta\left( \rho_{A_2}(\theta) - c_3 \right)}} \cdot \frac{1}{1 - e^{\theta\left( \rho_{A_2}(\theta) + \rho_{A_3}(\theta) - c_3 \right)}} \cdot \frac{1}{1 - e^{-\theta|c_1 - c_2|}} .$$

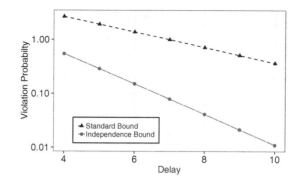

**Fig. 3.** Delay bound diamond network.

# 4 Numerical Evaluation

We present the results of a numerical evaluation for both case studies. We ran $10^4$ Monte-Carlo simulations to sample the parameters for different server rates and packet sizes, the latter sampled from an exponential distribution. The scenarios are then filtered to ensure a utilization $\in [0.5, 1)$.

## 4.1 Quality of the Bounds

**Diamond Network:** This topology, after above the mentioned filtering, yields 485 remaining scenarios, of which 371 are improved. The fact that not all are improved despite the avoidance of Hölder's inequality can be explained as follows: In the analysis, the Union bound is applied after Hölder's inequality. The exponentiation before the summing followed by a square root can have a "mitigating" effect. A similar observation has been exploited in SNC literature before [17].

We also measured the extent of the improvement by computing the ratio of the delay violation probability of the standard approach over the "independence bound". Clearly, values above 1 are desirable. Here, we obtain a median improvement of 6.04. In Fig. 3, we depict the delay bounds for specific parameters.

**The $\mathbb{L}$:** For this topology, we expect a weaker performance, as our approach using independence as a bound requires the additional step of extending the interval of one output process. The numerical results confirm this expectation: Out of the 729 scenarios, only half of them (384) yield a performance gain. The median of the improvement ratio confirms this, being relatively close to 1 (1.27). Again, we show the delay bounds for fixed parameters (Fig. 4).

## 4.2 Computation Run Time

Our proposed approach does not only often substantially improve the bounds but it also has a much lower computation complexity than the standard approach. The reason is that the latter relies on an additional Hölder parameter. The optimizations are conducted using a grid search followed by a downhill

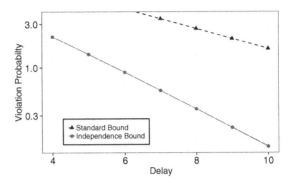

**Fig. 4.** Delay bound in the $\mathbb{L}$.

simplex algorithm. The improvements ratios are in the median 337.5 (1.62 sec compared to 0.0048 sec) for the diamond scenario and 458.1 for the $\mathbb{L}$ (1.42 s compared to 0.0031 s). These improvements due to the reduction of the optimization parameters indicates a significant potential for an analysis of larger networks, as the optimization step in the MGF-based SNC can severely limit its scalability.

## 5   Discussion

In this paper, we found interesting results indicating that by using independence as a bound, one can often times improve the delay bound while also speeding up the run time significantly. Obviously, the crucial next step is to find scenarios in which the conjecture can be proved rigorously. One potential technique might be to use the coupling method [15, 21], as it is can be applied to derive relations between tail probabilities. Furthermore, more scenarios can be analyzed in which the negative dependence can be exploited. In particular, this includes large-scale experiments that require many invocations of Hölder's inequality.

## A   Appendix

### A.1   Diamond Network

By using the conjecture, we have obtained so far that

$$P(d(t) > T)$$

$$\leq \sum_{\tau_1=0}^{t} \mathrm{E}\left[e^{\theta A_1(\tau_1,t)}\right] e^{-\theta c_1(t+T-\tau_1)} \mathrm{E}\left[e^{\theta\left(D_2^{(2)}+D_3^{(3)}\right)(\tau_1,t+T)}\right]$$

$$\leq \sum_{\tau_1=0}^{t} \mathrm{E}\left[e^{\theta A_1(\tau_1,t)}\right] e^{-\theta c_1(t+T-\tau_1)}$$

$$\times \mathrm{E}\left[e^{\theta\left((A_2\oslash[S_4-A_3]^+)\oslash S_2\right)(\tau_1,t+T)}\right] \mathrm{E}\left[e^{\theta((A_3\oslash S_4)\oslash S_3)(\tau_1,t+T)}\right].$$

This leads to

$$P(d(t) > T)$$

$$\leq \sum_{\tau_1=0}^{t} \mathrm{E}\left[e^{\theta A_1(\tau_1,t)}\right] e^{-\theta c_1(t+T-\tau_1)} \left\{ \sum_{\tau_2=0}^{\tau_1} \mathrm{E}\left[e^{\theta\left(A_2 \oslash [S_4 - A_3]^+\right)(\tau_2,t+T)}\right] \mathrm{E}\left[e^{-\theta S_2(\tau_2,\tau_1)}\right] \right\}$$

$$\cdot \left\{ \sum_{\tau_2=0}^{\tau_1} \mathrm{E}\left[e^{\theta(A_3 \oslash S_4)(\tau_2,t+T)}\right] \mathrm{E}\left[e^{-\theta S_3(\tau_2,\tau_1)}\right] \right\}$$

$$\leq \sum_{\tau_1=0}^{t} \mathrm{E}\left[e^{\theta A_1(\tau_1,t)}\right] e^{-\theta c_1(t+T-\tau_1)}$$

$$\cdot \left\{ \sum_{\tau_2=0}^{\tau_1} \left\{ \sum_{\tau_3=0}^{\tau_2} \mathrm{E}\left[e^{\theta A_2(\tau_3,t+T)}\right] \mathrm{E}\left[e^{\theta A_3(\tau_3,\tau_2)}\right] \mathrm{E}\left[e^{-\theta S_4(\tau_3,\tau_2)}\right] \right\} \mathrm{E}\left[e^{-\theta S_2(\tau_2,\tau_1)}\right] \right\}$$

$$\cdot \left\{ \sum_{\tau_2=0}^{\tau_1} \left\{ \sum_{\tau_3=0}^{\tau_2} \mathrm{E}\left[e^{\theta A_3(\tau_3,t+T)}\right] \mathrm{E}\left[e^{-\theta S_4(\tau_3,,\tau_2)}\right] \right\} \mathrm{E}\left[e^{-\theta S_3(\tau_2,\tau_1)}\right] \right\},$$

after applying the Union bound for each usage of the deconvolution. Further assuming all $A_i$ to be $(\sigma_A, \rho_A)$-bounded yields a closed-form for the delay bound under the stability condition

$$\rho_{A_1}(\theta) + \rho_{A_2}(\theta) + \rho_{A_3}(\theta) < c_1,$$
$$\rho_{A_2}(\theta) < c_2,$$
$$\rho_{A_3}(\theta) < c_3,$$
$$\rho_{A_2}(\theta) + \rho_{A_3}(\theta) < c_4 :$$

$$P(d(t) > T)$$

$$\overset{\text{(Definition 5)}}{\leq} \sum_{\tau_1=0}^{t} e^{\theta\left(\rho_{A_1}(\theta)(t-\tau_1)+\sigma_1(\theta)\right)} e^{-\theta c_1(t+T-\tau_1)}$$

$$\cdot \left\{ \sum_{\tau_2=0}^{\tau_1} \left\{ \sum_{\tau_3=0}^{\tau_2} e^{\theta\left(\rho_{A_2}(\theta)(t+T-\tau_3)+\sigma_{A_2}(\theta)\right)} e^{\theta\left(\rho_{A_3}(\theta)(\tau_2-\tau_3)+\sigma_{A_3}(\theta)\right)} e^{-\theta c_4(\tau_2-\tau_3)} \right\} e^{-\theta c_2(\tau_1-\tau_2)} \right\}$$

$$\cdot \left\{ \sum_{\tau_2=0}^{\tau_1} \left\{ \sum_{\tau_3=0}^{\tau_2} e^{\theta\left(\rho_{A_3}(\theta)(t+T-\tau_3)+\sigma_{A_3}(\theta)\right)} e^{-\theta c_4(\tau_2-\tau_3)} \right\} e^{-\theta c_3(\tau_1-\tau_2)} \right\}$$

$$\leq e^{\theta\left(\left(\rho_{A_2}(\theta)+\rho_{A_3}(\theta)-c_1\right)T+\sigma_1(\theta)+\sigma_{A_2}(\theta)+2\sigma_{A_3}(\theta)\right)}$$

$$\cdot \sum_{\tau_1=0}^{t} e^{\theta\left(\rho_{A_1}(\theta)-c_1\right)(t-\tau_1)} \left\{ \sum_{\tau_2=0}^{\tau_1} \frac{e^{\theta \rho_{A_2}(\theta)(t-\tau_2)}}{1 - e^{\theta\left(\rho_{A_2}(\theta)+\rho_{A_3}(\theta)-c_4\right)}} e^{-\theta c_2(\tau_1-\tau_2)} \right\}$$

$$\cdot \left\{ \sum_{\tau_2=0}^{\tau_1} \frac{e^{\theta \rho_{A_3}(\theta)(t-\tau_2)}}{1 - e^{\theta\left(\rho_{A_3}(\theta)-c_4\right)}} e^{-\theta c_3(\tau_1-\tau_2)} \right\}$$

$$\leq e^{\theta\left(\left(\rho_{A_2}(\theta)+\rho_{A_3}(\theta)-c_1\right)T+\sigma_1(\theta)+\sigma_{A_2}(\theta)+2\sigma_{A_3}(\theta)\right)}$$

$$\cdot \sum_{\tau_1=0}^{t} \frac{e^{\theta\left(\rho_{A_1}(\theta)+\rho_{A_2}(\theta)+\rho_{A_3}(\theta)-c_1\right)(t-\tau_1)}}{1 - e^{\theta\left(\rho_{A_2}(\theta)-c_2\right)}} \cdot \frac{1}{1 - e^{\theta\left(\rho_{A_3}(\theta)-c_3\right)}}$$

$$\cdot \frac{1}{1 - e^{\theta\left(\rho_{A_2}(\theta)+\rho_{A_3}(\theta)-c_4\right)}} \cdot \frac{1}{1 - e^{\theta\left(\rho_{A_3}(\theta)-c_4\right)}}$$

$$\leq \frac{e^{\theta\left(\left(\rho_{A_2}(\theta)+\rho_{A_3}(\theta)-c_1\right)T+\sigma_1(\theta)+\sigma_{A_2}(\theta)+2\sigma_{A_3}(\theta)\right)}}{1-e^{\theta\left(\rho_{A_1}(\theta)+\rho_{A_2}(\theta)+\rho_{A_3}(\theta)-c_1\right)}}\cdot\frac{1}{1-e^{\theta\left(\rho_{A_2}(\theta)-c_2\right)}}$$

$$\cdot\frac{1}{1-e^{\theta\left(\rho_{A_3}(\theta)-c_3\right)}}$$

$$\cdot\frac{1}{1-e^{\theta\left(\rho_{A_2}(\theta)+\rho_{A_3}(\theta)-c_4\right)}}\cdot\frac{1}{1-e^{\theta\left(\rho_{A_3}(\theta)-c_4\right)}},$$

where we used the convergence of the geometric series.

## A.2   The $\mathbb{L}$

We have that

$$P(d(t)>T)$$

$$\leq\sum_{\tau_1=0}^{t}\mathrm{E}\left[e^{\theta A_1(\tau_1,t)}\right]\sum_{\tau_2=\tau_1}^{t+T}e^{-\theta c_1\cdot(\tau_2-\tau_1)}e^{-\theta c_2\cdot(t+T-\tau_2)}\mathrm{E}\left[e^{\theta D_3^{(3)}(\tau_1,t+T)}e^{\theta D_2^{(3)}(\tau_1,\tau_2)}\right]$$

$$\leq\sum_{\tau_1=0}^{t}\mathrm{E}\left[e^{\theta A_1(\tau_1,t)}\right]\sum_{\tau_2=\tau_1}^{t+T}e^{-\theta c_1\cdot(\tau_2-\tau_1)}e^{-\theta c_2\cdot(t+T-\tau_2)}\mathrm{E}\left[e^{\theta D_3^{(3)}(\tau_1,t+T)}e^{\theta D_2^{(3)}(\tau_1,t+T)}\right].$$

With the conjecture, we compute

$$P(d(t)>T)$$

$$\leq\sum_{\tau_1=0}^{t}\mathrm{E}\left[e^{\theta A_1(\tau_1,t)}\right]\sum_{\tau_2=\tau_1}^{t+T}e^{-\theta c_1\cdot(\tau_2-\tau_1)}e^{-\theta c_2\cdot(t+T-\tau_2)}\mathrm{E}\left[e^{\theta D_3^{(3)}(\tau_1,t+T)}\right]\mathrm{E}\left[e^{\theta D_2^{(3)}(\tau_1,t+T)}\right]$$

$$\leq\sum_{\tau_1=0}^{t}\mathrm{E}\left[e^{\theta A_1(\tau_1,t)}\right]\sum_{\tau_2=\tau_1}^{t+T}e^{-\theta c_1\cdot(\tau_2-\tau_1)}e^{-\theta c_2\cdot(t+T-\tau_2)}\mathrm{E}\left[e^{\theta(A_2\oslash S_3)(\tau_1,t+T)}\right]$$

$$\mathrm{E}\left[e^{\theta\left(A_3\oslash[S_3-A_2]^+\right)(\tau_1,t+T)}\right]$$

$$\leq\sum_{\tau_1=0}^{t}\mathrm{E}\left[e^{\theta A_1(\tau_1,t)}\right]\sum_{\tau_2=\tau_1}^{t+T}e^{-\theta c_1\cdot(\tau_2-\tau_1)}e^{-\theta c_2\cdot(t+T-\tau_2)}\left\{\sum_{\tau_3=0}^{\tau_1}\mathrm{E}\left[e^{\theta A_2(\tau_3,t+T)}\right]e^{-\theta c_3(\tau_1-\tau_3)}\right\}$$

$$\cdot\left\{\sum_{\tau_3=0}^{\tau_1}\mathrm{E}\left[e^{\theta A_3(\tau_3,t+T)}\right]\mathrm{E}\left[e^{-\theta[S_3-A_2]^+(\tau_3,\tau_1)}\right]\right\}$$

$$\leq\sum_{\tau_1=0}^{t}\mathrm{E}\left[e^{\theta A_1(\tau_1,t)}\right]\sum_{\tau_2=\tau_1}^{t+T}e^{-\theta c_1\cdot(\tau_2-\tau_1)}e^{-\theta c_2\cdot(t+T-\tau_2)}\left\{\sum_{\tau_3=0}^{\tau_1}\mathrm{E}\left[e^{\theta A_2(\tau_3,t+T)}\right]e^{-\theta c_3(\tau_1-\tau_3)}\right\}$$

$$\cdot\left\{\sum_{\tau_3=0}^{\tau_1}\mathrm{E}\left[e^{\theta A_3(\tau_3,t+T)}\right]\mathrm{E}\left[e^{\theta A_2(\tau_3,\tau_1)}\right]e^{-\theta c_3(\tau_1-\tau_3)}\right\}.$$

If we again assume all $A_i$ to be $(\sigma_A,\rho_A)$-bounded, we obtain for

$$\rho_{A_1}(\theta)+\rho_{A_2}(\theta)+\rho_{A_3}(\theta)<\min\{c_1,c_2\},$$
$$\rho_{A_2}(\theta)+\rho_{A_3}(\theta)<c_3,$$

and $c_1 \neq c_2$:

$$\mathrm{P}(d(t) > T)$$

(Definition 5)
$$\leq \sum_{\tau_1=0}^{t} e^{\theta\left(\rho_{A_1}(\theta)(t-\tau_1)+\sigma_{A_1}(\theta)\right)} \sum_{\tau_2=\tau_1}^{t+T} e^{-\theta c_1 \cdot (\tau_2-\tau_1)} e^{-\theta c_2 \cdot (t+T-\tau_2)}$$

$$\cdot \left\{ \sum_{\tau_3=0}^{\tau_1} e^{\theta\left(\rho_{A_2}(\theta)(t+T-\tau_3)+\sigma_{A_2}(\theta)\right)} e^{-\theta c_3 (\tau_1-\tau_3)} \right\}$$

$$\cdot \left\{ \sum_{\tau_3=0}^{\tau_1} e^{\theta\left(\rho_{A_2}(\theta)(\tau_1-\tau_3)+\sigma_{A_2}(\theta)\right)} e^{\theta\left(\rho_{A_3}(\theta)(t+T-\tau_3)+\sigma_{A_3}(\theta)\right)} e^{-\theta c_3 (\tau_1-\tau_3)} \right\}$$

$$\leq e^{\theta\left(\left(\rho_{A_2}(\theta)+\rho_{A_3}(\theta)\right) \cdot T+\sigma_{A_1}(\theta)+2\sigma_{A_2}(\theta)+\sigma_{A_3}(\theta)\right)}$$

$$\cdot \sum_{\tau_1=0}^{t} e^{\theta\left(\rho_{A_1}(\theta)+\rho_{A_2}(\theta)+\rho_{A_3}(\theta)\right)(t-\tau_1)} \sum_{\tau_2=\tau_1}^{t+T} e^{-\theta c_1 \cdot (\tau_2-\tau_1)} e^{-\theta c_2 \cdot (t+T-\tau_2)}$$

$$\cdot \left\{ \sum_{\tau_3=0}^{\tau_1} e^{\theta\left(\rho_{A_2}(\theta)-c_3\right)(\tau_1-\tau_3)} \right\} \left\{ \sum_{\tau_3=0}^{\tau_1} e^{\theta\left(\rho_{A_2}(\theta)+\rho_{A_3}(\theta)-c_3\right)(\tau_1-\tau_3)} \right\}$$

$$\leq e^{\theta\left(\left(\rho_{A_2}(\theta)+\rho_{A_3}(\theta)\right) \cdot T+\sigma_{A_1}(\theta)+2\sigma_{A_2}(\theta)+\sigma_{A_3}(\theta)\right)}$$

$$\cdot \sum_{\tau_1=0}^{t} \frac{e^{\theta\left(\rho_{A_1}(\theta)+\rho_{A_2}(\theta)+\rho_{A_3}(\theta)\right)(t-\tau_1)}}{1-e^{\theta\left(\rho_{A_2}(\theta)-c_3\right)}} \cdot \frac{1}{1-e^{\theta\left(\rho_{A_2}(\theta)+\rho_{A_3}(\theta)-c_3\right)}}$$

$$\cdot \sum_{\tau_2=\tau_1}^{t+T} e^{-\theta c_1 \cdot (\tau_2-\tau_1)} e^{-\theta c_2 \cdot (t+T-\tau_2)}$$

$$\leq e^{\theta\left(\left(\rho_{A_2}(\theta)+\rho_{A_3}(\theta)\right) \cdot T+\sigma_{A_1}(\theta)+2\sigma_{A_2}(\theta)+\sigma_{A_3}(\theta)\right)}$$

$$\cdot \sum_{\tau_1=0}^{t} \frac{e^{\theta\left(\rho_{A_1}(\theta)+\rho_{A_2}(\theta)+\rho_{A_3}(\theta)\right)(t-\tau_1)}}{1-e^{\theta\left(\rho_{A_2}(\theta)-c_3\right)}} \cdot \frac{1}{1-e^{\theta\left(\rho_{A_2}(\theta)+\rho_{A_3}(\theta)-c_3\right)}}$$

$$\cdot \sum_{\tau_2=\tau_1}^{t+T} e^{-\theta c_1 \cdot (\tau_2-\tau_1)} e^{-\theta c_2 \cdot (t+T-\tau_2)}$$

$$\leq \frac{e^{\theta\left(\left(\rho_{A_2}(\theta)+\rho_{A_3}(\theta)-\min\{c_1,c_2\}\right) \cdot T+\sigma_{A_1}(\theta)+2\sigma_{A_2}(\theta)+\sigma_{A_3}(\theta)\right)}}{1-e^{\theta\left(\rho_{A_1}(\theta)+\rho_{A_2}(\theta)+\rho_{A_3}(\theta)-\min\{c_1,c_2\}\right)}}$$

$$\cdot \frac{1}{1-e^{\theta\left(\rho_{A_2}(\theta)-c_3\right)}} \cdot \frac{1}{1-e^{\theta\left(\rho_{A_2}(\theta)+\rho_{A_3}(\theta)-c_3\right)}} \cdot \frac{1}{1-e^{-\theta|c_1-c_2|}},$$

where we used again the convergence of the geometric series.

## References

1. Antonini, R.G., Kozachenko, Y., Volodin, A.: Convergence of series of dependent $\varphi$-subgaussian random variables. J. Math. Anal **338**, 1188–1203 (2008)

2. Baccelli, F., Cohen, G., Olsder, G.J., Quadrat, J.P.: Synchronization and Linearity: An Algebra for Discrete Event Systems. Wiley, Chichester (1992)
3. Beck, M.A.: Advances in Theory and Applicability of Stochastic Network Calculus. Ph.D. thesis, TU Kaiserslautern (2016)
4. Chang, C.S.: Performance Guarantees in Communication Networks. Springer, London (2000). https://doi.org/10.1007/978-1-4471-0459-9
5. Ciucu, F., Burchard, A., Liebeherr, J.: Scaling properties of statistical end-to-end bounds in the network calculus. IEEE/ACM ToN 52(6), 2300–2312 (2006)
6. Cruz, R.L.: A calculus for network delay, part I: network elements in isolation. IEEE Trans. Inf. Theor. 37(1), 114–131 (1991)
7. Cruz, R.L.: A calculus for network delay, part II: network analysis. IEEE Trans. Inf. Theor. 37(1), 132–141 (1991)
8. Cruz, R.L.: Quality of service management in integrated services networks. In: Proceedings of Semi-Annual Research Review, CWC, vol. 1, pp. 4–5 (1996)
9. Dong, F., Wu, K., Srinivasan, V.: Copula analysis for statistical network calculus. In: Proceedings of the IEEE INFOCOM'15, pp. 1535–1543 (2015)
10. Dubhashi, D., Ranjan, D.: Balls and bins: a study in negative dependence. Random Struct. Algor. 13(2), 99–124 (1998)
11. Fidler, M.: An end-to-end probabilistic network calculus with moment generating functions. In: Proceedings of the IEEE IWQoS'06, pp. 261–270 (2006)
12. Jiang, Y., Liu, Y.: Stochastic Network Calculus, vol. 1, p. XIX, 232. Springer, London (2008)
13. Joag-Dev, K., Proschan, F.: Negative association of random variables with applications. Ann. Stat. 11(1), 286–295 (1983)
14. Lehmann, E.L.: Some concepts of dependence. Ann. Math. Stat. 37(5), 1137–1153 (1966)
15. Lindvall, T.: Lectures on the Coupling Method. Courier Corporation (2002)
16. Nelson, R.: Probability, Stochastic Processes, and Queueing Theory: The Mathematics of Computer Performance Modeling. Springer, New York (1995). https://doi.org/10.1007/978-1-4757-2426-4
17. Nikolaus, P., Schmitt, J., Schütze, M.: h-Mitigators: improving your stochastic network calculus output bounds. Comput. Commun. 144, 188–197 (2019)
18. Rizk, A., Fidler, M.: Leveraging statistical multiplexing gains in single-and multi-hop networks. In: Proceedings of the IEEE IWQoS'11, pp. 1–9 (2011)
19. Schmitt, J., Zdarsky, F.A., Fidler, M.: Delay bounds under arbitrary multiplexing: When network calculus leaves you in the lurch ... In: Procedings of the IEEE International Conference on Computer Communications (INFOCOM'08). Phoenix, AZ, USA (2008)
20. Sung, S.H.: On the exponential inequalities for negatively dependent random variables. J. Math. Anal. Appl. 381(2), 538–545 (2011)
21. Thorisson, H.: Coupling. Stationarity and Regeneration. Springer, New York (2000)

# Map-Reduce Process Algebra: A Formalism to Describe Directed Acyclic Graph Task-Based Jobs in Parallel Environments

Enrico Barbierato[1], Marco Gribaudo[2(✉)], and Mauro Iacono[3]

[1] Università Cattolica del Sacro Cuore, via dei Musei 41, 25121 Brescia, Italy
`enrico.barbierato@unicatt.it`
[2] Dipartimento di Elettronica, Informazione e Bioingegneria, Politecnico di Milano, via Ponzio 345, 20133 Milan, Italy
`marco.gribaudo@polimi.it`
[3] Dipartimento di Matematica e Fisica, Università degli Studi della Campania "L. Vanvitelli", viale Lincoln 5, 81100 Caserta, Italy
`mauro.iacono@unicampania.it`

**Abstract.** Cloud Computing has made possible flexible resources provisioning from an almost unlimited pool. This has created the opportunity to broaden the horizon of data that can be analyzed, allowing to support the so called Big Data Analytics applications. New programming paradigms, such as NoSQL queries and Map-Reduce applications, have emerged within frameworks such as Microsoft Azure, Hadoop and Apache Spark. In many cases, applications execute jobs that are split into stages, each one composed of tasks that can be run in parallel on many computational nodes. Directed acyclic graphs describe the precedence between stages, defining the execution rules and controlling the degree of parallelism. This work presents a Process Algebra dialect aimed at describing both jobs and execution environments. The proposed framework is then used to model and study standard parallel programming benchmarks, to demonstrate its applicability.

**Keywords:** Process algebra · Fork-join · Map-Reduce · dagSim

## 1 Introduction

Big Data computing is definitely here to stay. Far from being a fashion trend, the need for Big Data applications and analytics is creating one of the most profitable and competence markets. Competitors work to enhance the power of bare metal in huge data centers with proper architectural models and paradigms running for leadership. Fostered by the big Internet companies, tools have been built around this motivation. Based on automation of workloads parallelization, they boosted to a new scale, previously unseen in commercial architectures.

© Springer Nature Switzerland AG 2020
M. Gribaudo et al. (Eds.): ASMTA 2019, LNCS 12023, pp. 85–99, 2020.
https://doi.org/10.1007/978-3-030-62885-7_7

A foundation on simple highly scalable software paradigms such as Map-Reduce, was the key to leverage large computing facilities to harvest a massive, variegate data.

Besides powerful software frameworks, such as Microsoft Azure, Hadoop, Apache Spark, and data layers such as NoSQL based solutions or analogous file-system-like data storage and management substrates, a proper control on operations and resources is needed to stay upfront and keep pushing profitability and computational efficiency. Events monitoring in datacenters represents the main tool to accomplish many purposes. Specifically, it enables a continuous management of the infrastructure and a dynamic tuning of resource allocation. Furthermore, it optimizes performances and power consumption, and keeps the Total Cost of Ownership (TCO) as low as possible with respect to the high variability of the workloads.

## 1.1  Big Data Applications Performance Prediction

Tracing system events and processing them to extract a solid knowledge about the history of the overall infrastructure and of its parts can provide a precious support for administrators, who need a view on the system that goes beyond a glance on its dynamics to be able to take the right decisions. Given the scale of the system, every approximation in the estimation of system performance related metrics, or of the effects of the main factors that can influence them, may result in a significant fluctuation in costs. While system traces may help building an accurate model of the past to understand hidden and emerging phenomena, they are less reliable when there is a requirement for planning future resource use and allocation. In this case, predictive models are needed to improve available information about the current and past state of the system with an abstraction of future behavior and interactions between workloads.

Knowledge about future workloads of large computing systems serving multiple applications and users is generally not available, or poorly approximated. Most of the workload of Big Data applications consists of the part of computation that can be automatically spawned on a large number of processors. This approach obeys the logic of the underlying computing paradigm, e.g. Map-Reduce. This aspect greatly helps shaping significant performance models that allow to anticipate the needs and the dynamics of the system. The authors previously dealt with the problem of performance prediction in large scale Big Data applications studying the effects of storage subsystems [6], NoSQL subsystems [10] (including Hadoop influences), frameworks [8] (focusing on Apache Hive). This work proposes an approach that leverages the structure of Apache Spark oriented workloads to produce a specific description of them that is suitable for simulation.

## 1.2  Map-Reduce Process Algebra (MRPA)

The description is based on a formalism, namely Map-Reduce Process Algebra (MRPA), specifically designed for this purpose and presented in this paper for the first time. This approach leverages previous experiences [9] and is meant to be integrated in the SIMTHESys Modeling Framework [7,11], together with the dagSIM simulator, previously developed by the authors and described in details in the next section. MRPA is a modeling language (or formalism, in the SIMTHESys terminology). It is inspired to Process Algebra semantics (especially Performance Evaluation Process Algebra, or PEPA) and builds its primitives to represent Apache Spark based applications. The dagSIM simulator has been chosen for its orientation towards event based simulation of Directed Acyclic Graphs (DAG) represented workloads, suitable to capture the structure of Apache Spark based Big Data applications. As the dagSIM simulator was not meant to support the complexity of MRPA models, the SIMTHESys approach has been exploited to generate a solver that can generate dagSIM compatible models from MRPA models, applying a lightweight refactoring process to dagSIM by means of SIMTHESys (the description of this process is out of the scope of this paper). Consequently, even if not on the same semantic level, Apache Spark, Process Algebra, PEPA and Fork-Join can be considered the four pillars of this approach.

### 1.3   Originality of the Contribution

The original contribution of the paper mainly consists of the semantic definition of a new, compact Process Algebra dialect operator implementing the Fork-Join paradigm called MRPA. Specifically, this operator describes a scenario where a new stage starts invoking a set of tasks, operating in parallel and terminating one by one. When the execution of all the tasks has come to an end, the next stage in the DAG modeling the query can start. With respect to earlier work, the novelty of the approach described in this paper encompasses the following characteristics: the operator must be i) timed, in order to properly evaluate the performance and ii) semantically close to Process Algebra's formalisms to exploit the existing mechanisms and leverage its capability to model real systems.

The paper is structured as follows: after this introduction, Sect. 2 discusses in details the four pillars; Sect. 3 describes the MRPA formalism and defines its semantics; Sect. 4 proposes an application to show the effectiveness of the approach; Sect. 5 presents related work in literature; finally, Sect. 6 draws conclusions and introduces future developments planned for this research.

## 2   The Four Pillars of the Approach

This section provides a short introduction to the background and the concepts used to support the developed formalism, specifically the foundation of Apache Spark (2.1), Process Algebra (2.2), PEPA (2.3) and the fork-join paradigm (2.4).

### 2.1   Apache Spark

Apache Spark is a in-memory data analytics system, meaning that data are processed in the main memory to avoid the unnecessary I/O interactions with the storage units. Besides being distributed, one of its most important properties is to be highly scalable and versatile, as it is possible to develop applications in most of well known programming languages such as Java, Python and R. This is made possible by the fact that Spark is essentially a Java Virtual Machine.

Nowadays, Apache Spark is generally regarded as the inheritor of Map-Reduce method. In this sense, a Spark application replicates the structure of its predecessor, which consists of a self-contained computation executing user-supplied code. In Spark, the computation unit at the top level privileges an application. Runtime Spark applications map to a single *driver* process and a number of *executor* processes distributed across the hosts contained in a cluster. A *driver* has the responsibility of managing the job flow. It behaves also as a task scheduler component, being available until the application is running. An *executor* literally executes the requested work, which is in turn decomposed into a set of *tasks*. An executor can be allocated both statically or dynamically.

There is correspondence between the concept of *action* and *job*, as each time a Spark application detects that an *action* has been invoked, it launches a corresponding *job*. At a higher level, the action is related to a dataset: this information is used by Spark in order to determine how the execution can be planned, in order to merge the dataset transformation into a set of *stages*. Ultimately, a *stage* is a tasks agglomeration executing identical code on the executors, each one on different data subset. A *job* is a collection of *stages*.

To perform data processing in an fast and efficient way, Spark exploits an engine based on DAGs. When Spark is invoked, it converts the requested application into a DAG composed of tasks.

## 2.2   Process Algebra

The idea behind a Process Algebra (PA) is based on two concepts: i) the abstraction of the behavior of a generic system (software, mechanical, etc.) with the term *process* on one side and ii) to describe a process by using an algebraic process based on axioms, aiming to perform a kind of computation. In a way, a PA shares some of its concepts with an automaton, which is described by states and transitions. Unfortunately, automata are limited since the notion of interaction is not taken in account.

Historically, PA stemmed from the Calculus of Communicating Systems (CCS), depicted by Milner in 1980 [21]. *Agents* and *Actions* represent the key concepts. An *action* can be either observable or not, while an *Agent* can be further subdivided into input and output actions. An *Agent* represents a process and can be defined by one of the following atomic *actions*.

*Action Prefixing.* Given an *Action a* and a process P, then a.P is a process, meaning that P is active only when the *a* has been completed, representing a sequential composition.

*Choice Operator.* Given two processes, $X$, $Y$, then $X + Y$ is a process executing $X$ or $Y$ in a non-deterministic way.

*Parallel Composition.* Given two not communicating processes, $X$, $Y$, then $X \parallel Y$ is a process where $X$ and $Y$ are executed in parallel.

*Communication.* Given two processes, $X$, $Y$, then the system composed of $X \mid Y$ is denoted by the fact that X and Y work independently and communicate via complementary ports.

*Restriction.* Given a process $X$ and a set of actions $\Sigma$, then $X \, . \, \Sigma$ is a process that cannot execute actions in the set $\Sigma$.

Time is not considered in CCS, therefore all the actions are instantaneous. Choices are non deterministic.

## 2.3   PEPA

Performance Evaluation of Process Algebra (PEPA, see [18], available at http://www.dcs.ed.ac.uk/pepa/) is a formal, textual language based on standard PA to model distributed systems. With respect to the classic PA, it includes timing properties. A PEPA model consists of a set of composed components. Each one can perform a single activity or cooperates on those that are shared. The timing information allows the modeler to specify a rate (a negative random variable exponentially distributed) by which the activity must be completed.

An activity $a = (\alpha, r)$ is denoted by i) a label $\alpha$ and ii) a positive real number $r$, i.e. the parameter of an exponentially distributed random variable describing a delay. PEPA exploits a set of operators to define its semantics as follows.

*Prefix.* A *behavior* is described by the Prefix operator. Component $(\alpha, \text{r}).P$ accomplishes the activity $(\alpha, \text{r})$, acting as $P$.

*Choice.* This operator applies to the activities of $P$ and $Q$ in the form $P + Q$. It denotes a system behaving as only one of $P$, $Q$.

*Cooperation.* Given a *cooperation set* $L$, $P$ and $Q$ are executed independently and in parallel with any activity whose type $\alpha \notin L$. When this condition is violated, then the components are bound to cooperate in what is called *shared activity*. This operator is indicated with the $\bowtie_{L}$ notation. If $L = \emptyset$, the short-hand notation $\parallel$ is preferred.

*Hiding.* The purpose of this operator is literally to hide some aspect of the component behavior in order to perform a degree of abstraction. Given a set $L$, then $P \, / \, L$ reflects the same behavior of $P$ with exception of all the activities $a \in L$ are hidden.

*Constant.* A constant is a component. This operator allows a constant to behave like a component. For example, the syntax $A \overset{\text{def}}{=} P$ forces $A$ behaving like $P$ component.

## 2.4   The Fork-Join Paradigm

The *fork-join* paradigm exploits a parallelism operator on the basis of the *divide and conquer* principle. Coherently with this approach, a system can be decomposed into smaller independent components (i.e. they are not bound to perform the same execution) that are solved in a parallel fashion. The decomposition granularity cannot be arbitrary though, as scheduling overhead issues may arise. Typically, a *fork-join* operation is composed of three constructs: i) *Fork*, whose purpose is to initiate an independent flow of executions, ii) *Quit*, which ends the execution of the flow and iii) *Join* operation, which is responsible of merging flow control once all the subtasks have completed their execution.

Although both Map-Reduce and Fork-Join are general techniques and have been applied in multiple contexts and compute systems, the latter approach presents some specific differences (see [23], where the authors introduce also a hybrid formalism). The latter has been designed to operate on a Java Virtual Machine and not on large clusters. The idea is to partition a process into sub-processes being therefore rather different from Map-Reduce technique, since the latter performs one partition only, without any communication between the partitions until the *reduce* phase. The application target is also different, being multi-cores platform for *fork-join* and cloud-based scenarios for Map-Reduce.

## 3   The Map-Reduce Process Algebra (MRPA) Formalism

Based on [17], the syntax of MRPA can be formally introduced by means of the following grammar:

$$S ::= (\alpha, r).S | S + S | T.S | C_S$$
$$T ::= (\alpha, r) | T.T | T + T | T\&T | C_T$$
$$P ::= P \underset{L}{\bowtie} P | P/L | C_S$$

The main addition is the introduction of tasks $T$. A *Task* is a component which might also end. It is characterised by timed actions, sequence and choice as regular components, but it can only continue in other tasks, and it might also *end* - reach an action that does not continue in a constant. Tasks can also use the *Map-Reduce* operator $\&$. The key idea of that operator, is that it splits a task $T$ into two tasks $T_1$ & $T_2$. The task can the continue (or end) only when both $T_1$ and $T_2$ have ended.

Sequential components are then enriched by operator $T.S$: the components evolves as task $T$ and then continues as $S$ as soon as task $T$ has ended. We then use the following short-hand notation:

$$T ::= n \times T, \quad n \in \mathbb{N} \quad \equiv \quad T ::= \overbrace{T\&T\ldots\&T}^{n}$$

to denote a task $T$ that forks itself into $n \in \mathbb{N}$ subtasks. Constants for sequential and task components can only belong to the corresponding type: we thus use $C_S$ to denote sequential components only, $C_T$ task components.

In this framework, a classical Map-Reduce job composed of respectively of $N_{Map}$ and $N_{Red}$ tasks, submitted by $N_{users}$ users, and executed on $N_{nodes}$ nodes, can be modelled as follows:

$$Job \stackrel{def}{=} (think, Z).Map.Reduce.Job$$

$$Map \stackrel{def}{=} N_{Map} \times ((get, \lambda_{Map}).(rel, \mu_{Map}))$$

$$Reduce \stackrel{def}{=} N_{Red} \times ((get, \lambda_{Red}).(rel, \mu_{Red}))$$

$$Node \stackrel{def}{=} (get, \top).(rel, \top).Node$$

$$Sys \stackrel{def}{=} \underbrace{(Job \,||\, \cdots \,||\, Job)}_{N_{users}} \underset{(get,rel)}{\bowtie} \underbrace{(Node \,||\, \cdots \,||\, Node)}_{N_{nodes}}$$

where $Z$ is the think time (the time a user spends before submitting the new job), $\lambda_{Map}$ and $\lambda_{Red}$ are the task execution speed values for the two stages, and $\mu_{Map}$ and $\mu_{Red}$ represent the rate at which resources are released. Note that since in the end these type of models will be solved via discrete event simulation, the exponential firing time assumption can be relaxed. To further simplify the notation, the classical population model notation is used:

$$P[n] \equiv \underbrace{P \,||\, \cdots \,||\, P}_{n} \tag{1}$$

In this way, the system definition becomes:

$$Sys \stackrel{def}{=} Job[N_{users}] \underset{(get,rel)}{\bowtie} Node[N_{nodes}] \tag{2}$$

## 3.1   DAG Examples

A simple Map-Reduce job, whose DAG is represented in Fig. 1(a), can be described in MRPA. However, more complex Spark jobs can be modelled with the proposed formalism. In particular, all settings include the following definitions:

$$M_i \stackrel{def}{=} N_{Map_i} \times ((get, \lambda_{Map_i}).(rel, \mu_{Map_i}))$$

$$R_i \stackrel{def}{=} N_{Red_i} \times ((get, \lambda_{Red_i}).(rel, \mu_{Red_i}))$$

$$Node \stackrel{def}{=} (get, \top).(rel, \top).Node$$

$$Sys \stackrel{def}{=} Job[N_{users}] \underset{(get,rel)}{\bowtie} Node[N_{nodes}]$$

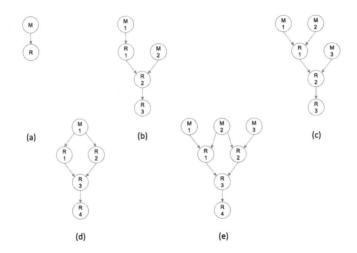

**Fig. 1.** Five typical Spark queries and their corresponding DAGs

Only the structure of the *Job* component is modified. For example, the simple Map-Reduce presented in Fig. 1(a) is characterised by:

$$Job \stackrel{def}{=} (think, Z).M_1.R_1.Job \tag{3}$$

Let us focus on the DAG presented in Fig. 1(b). In this case the Spark job is composed of 5 stages: two maps, and three reduces. In particular, reduce $R_1$ occurs after map $M_1$; reduce $R_2$ can only occur after $R_1$, but it also requires results from map $M_2$. A final reduce operation $R_3$ concludes the job. Since $M_1$ and $M_2$ have no previous dependencies, they can start immediately. However, $R_2$ can only start after $M_1$, $M_2$ and $R_1$ have completed. The corresponding job definition in MRPA is the following:

$$Job \stackrel{def}{=} (think, Z).((M_1.R_1)\&M_2).R_2.R_3.Job \tag{4}$$

Note how the term $((M_1.R_1)\&M_2)$ corresponds to a task definition. Since $M_1.R_1$ and $M_2$ work in parallel, they can potentially start at the same time (if they can acquire a *Node* resource). However, both $M_1.R_1$ and $M_2$ should have completed their execution to allow the entire task to move to $R_2$.

The DAG shown in Fig. 1(c), adds a map stage $M_3$, and requires an extra precedence, since now $R_1$ depends on both $M_1$ and $M_2$. The corresponding MRPA component is:

$$Job \stackrel{def}{=} (think, Z).(((M_1\&M_2).R_1)\&M_3).R_2.R_3.Job \tag{5}$$

The extra dependency is caught by adding an extra parallelism to the term $((M_1\&M_2).R_1)\&M_3$.

Figure 1(d) replaces the second map $M_2$ of Fig. 1(b) with another reduce stage. In this case, the completion of the only map stage $M_1$ enables both the execution of $R_1$ and $R_2$, which can be both executed in parallel, and needs to be completed to allow the job to continue to $M_3$. The MRPA specification for this Spark job is the following:

$$Job \stackrel{def}{=} (think, Z).M_1.(R_1 \& R_2).R_3.R_4.Job \qquad (6)$$

Note how this common dependence is simply obtained by exploiting the prefix operator before the Map-Reduce element.

The last DAG, shown in Fig. 1(e), is characterised by more complex dependencies. In particular, it has three map stages $M_1$, $M_2$ and $M_3$ which can immediately start. The next two reduce stages however have a common dependency on $M_2$, and $R_1$ can start if also $M_1$ has completed, while $R_2$ requires $M_3$ to be executed. This behaviour can be obtained by exploiting the properties of Process Algebra, with the following model:

$$Job \stackrel{def}{=} (think, Z).(M_1 \& M_2 \& M_3 \& R_1 \& R_2).R_3.R_4.Job$$

$$M_i \stackrel{def}{=} (N_{Map_i} \times ((get, \lambda_{Map_i}).(rel, \mu_{Map_i}))).(Mex_i, \infty) \quad (i = 1, 2, 3)$$

$$R_1 \stackrel{def}{=} (Mey_1, \infty).(Mey_2, \infty).(N_{Red_1} \times ((get, \lambda_{Red_1}).(rel, \mu_{Red_1})))$$

$$R_2 \stackrel{def}{=} (Mey_2, \infty).(Mey_3, \infty).(N_{Red_2} \times ((get, \lambda_{Red_2}).(rel, \mu_{Red_2})))$$

$$R_i \stackrel{def}{=} N_{Red_i} \times ((get, \lambda_{Red_i}).(rel, \mu_{Red_i})) \quad (i = 3, 4)$$

$$Mx_i \stackrel{def}{=} (Mex_i, \top).My_i \quad (i = 1, 2, 3)$$

$$My_i \stackrel{def}{=} (Mey_i, \top).My_i + (think, \top).Mx_i \quad (i = 1, 2, 3)$$

$$Node \stackrel{def}{=} (get, \top).(rel, \top).Node$$

$$Sys \stackrel{def}{=} Mx_i \underset{(Mex_i, Mey_i, think)}{\bowtie} Job[N_{users}] \underset{(get, rel)}{\bowtie} Node[N_{nodes}]$$

Please note that the infinity symbol indicates the rate for immediate actions. The new components $Mx_i$ and $My_i$ act as flags that check whether stage $M_i$ has completed. In particular, when stage $M_i$ ends, it synchronises with $Mx_i$ through action $Mex_i$, leading to $My_i$. Stages $R_1$ and $R_2$ can start only if they synchronise first with the required $Mey_i$ actions. The flags are reset when a new job starts, thank to the *think* action. Since this event does not require time, we have used the notation $\infty$ to specify that the corresponding event relates to an immediate action (that is, with its rate tending to $\infty$). Finally, notation $Mx_i \underset{(Mex_i, Mey_i, think)}{\bowtie}$ is used to define that components $Mx_1$, $Mx_2$ and $Mx_3$ synchronise with the corresponding action of the jobs components. Due to this explicit synchronisation, tasks $M_1$, $M_2$, $M_3$, $R_1$ and $R_2$ start at the same time. Synchronisation with the $Mx_i$ process occurs only once per $M_i$ component, at the end of the stage. The synchronisation of the components $My_1$ and $My_3$ with the reduce stages $R_1$ and $R_2$ occurs only once at their beginning. Synchronisation with $My_2$ occurs instead two times: one with $R_1$ and one with $R_2$.

## 3.2   Solution with the dagSim Simulator

To solve the models, the analysis is restricted to models from which a DAG compatible with the ddagSim tool specification format can be extracted, leaving the realisations of a tool fully supporting MRPA for future work. Details of this conversion process are outside the scope of this work. dagSim [3] is a high speed discrete event simulator oriented to DAGs analysis modeling Map-Reduce and Spark jobs[1]. In particular, we convert MRPA components to the equivalent dagSim models, and use the tool to study them.

A dagSim model is built according on the following entities: i) the dag stages and ii) the foreseen workload. In this context, a dag is defined as

$$\mathrm{DAG} = \left(S, N_{\mathrm{nodes}}, N_{\mathrm{users}}, \mathcal{Z}\right),\tag{7}$$

where $N_{\mathrm{nodes}} \in \mathbb{N}, N_{\mathrm{nodes}} \geq 1$ represents the number of computational nodes $N_{\mathrm{users}} \in \mathbb{N}, N_{\mathrm{users}} \geq 1$ the number of users concurrently submitting jobs to the system, and $\mathcal{Z}$ is the "think time distribution", i.e., the time a user will wait before submitting a new job. Set $S = \left\{s_1, \ldots, s_{N_{\mathrm{Stages}}}\right\}$ is the set of stages that define the DAG. Furthermore, each stage $s_i \in S$ is a tuple:

$$s_i = \left(\mathtt{id}, N_{\mathrm{Tasks}}, Pre, Post, \mathcal{T}\right),\tag{8}$$

where id is a symbolic constant assigning a name to the stage, $N_{\mathrm{Tasks}} \in \mathbb{N}, N_{\mathrm{Tasks}} \geq 1$ accounts for the tasks composing the stage, $Pre \in S$ and $Post \in S$ define respectively the stages that must have been completed for $s_i$ to be executable, and the set of stages that will be able to run after the completion of $s_i$. Finally, $\mathcal{T}$ represents the task duration distribution for the considered stage.

dagSim simulator is built by following the architecture of a classic discrete event simulation algorithm. Further enhancement have been implemented in order to achieve high performance. The current market witnesses different competitors at commercial level such as Arena (https://www.arenasimulation.com). In this respect, though dagSim is a lightweight tool, it aims specifically at DAG-based models. Furthermore, its proprietary scheduler library (see the manual at https://github.com/eubr-bigsea/dagSim/blob/master/simlib/Documentation/scheduler/manual/1.63/manual.html) exploits on one side optimized data structures (such as lists, heaps or calendar queue), on the other an engine written in C and LUA language allowing high speed processing even when the DAG's simulation generates a high number of events. The tool does not require external tools or libraries, therefore it is portable.

By using a doubly-linked list storing the relevant information about the tasks belonging to stages for which all dependencies have been satisfied and thus can start, it is possible to determine which one can be executed without performing a full search on the complete set of tasks in the DAG. From this point of view, the technique provided by dagSim's engine is original and more efficient with

---

[1] The tool is available at https://github.com/eubr-bigsea/dagSim. Extract the compressed package in a folder and compile dagSim by using the"make" utility. The usage is the following: ./dagSim < configurationfile.lua.

respect to other scheduling mechanisms implemented in general purpose tools such as JMT [12] or GreatSPN [14].

## 4  Experiments

To show the potentialities of the proposed approach, the models corresponding to the DAGs shown in Fig. 1(b), (c) and (d) are studied by using the parameters given in Table 1. It is assumed that the release time is negligible (that is $\mu_{Map_i} = \mu_{Red_i} = \infty$ for all stages $i$), $N_{users} = 3$ users, and a think time $\frac{1}{Z} = 20$ s.

**Table 1.** Experiment parameters

| Stage | $N_{task}$ | $\frac{1}{\lambda_{task}}$ | Stage | $N_{task}$ | $\frac{1}{\lambda_{task}}$ |
|---|---|---|---|---|---|
| $M_1$ | 100 | 100 ms | $R_1$ | 150 | 5 ms |
| $M_2$ | 80 | 10 ms | $R_2$ | 20 | 30 ms |
| $M_3$ | 100 | 20 ms | $R_3$ | 10 | 600 ms |
| | | | $R_4$ | 400 | 80 ms |

The job completion time distribution is regarded as a function of the available nodes $N_{nodes}$. The 95% confidence interval is reported in Fig. 2. However, due to the high accuracy of the simulation, upper and lower bounds cannot be distinguished from the average. As it can be seen, the additional component of case (c) has an impact on the performance only when the number of nodes is limited. Then, execution of the extra stage is embedded in the time resources would be idle waiting for the next phase to start, making cases (b) and (c) identical. In both cases, it is clear that the advantages of having $N_{nodes} > 30$ is minimal, and for $N_{nodes} > 40$ negligible. This shows the importance of being able to simulate complex DAG based jobs, to avoid over-provisioning the computational resources. The presence of an extra stage with a large number of tasks in case (d) (stage $R_4$), increases the response time, but makes performance gains still appreciable up to $N_{nodes} > 90$.

The capability of computing not only average values represents another advantage of using the dagSim simulator. Figure 3 shows the response time distribution of the job completion time for the three considered cases, with either $N_{nodes} = 20$ or $N_{nodes} = 30$. It is interesting to see that the difference between cases (b) and (c) tends to become minimal and less dependent on the number of nodes as the percentile increases, making the 98% percentile almost identical for both cases and resource configurations. Even if less evident, this is also visible in case (d). In other words, by increasing the percentile of the response time, if a constraint can be satisfied, it becomes less influenced by the system configuration.

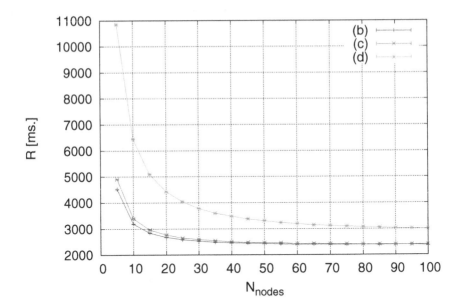

**Fig. 2.** Average response time for different number of nodes $N_{nodes}$, for the DAGs shown in Fig. 1(b), (c) and (d).

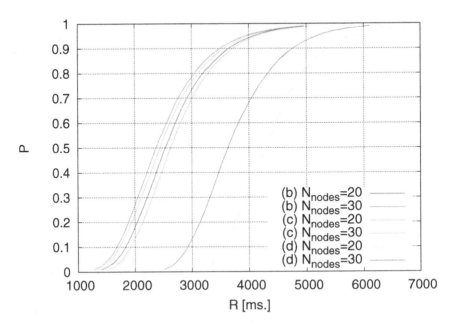

**Fig. 3.** Response time distribution for different number of nodes $N_{nodes}$, for the DAGs shown in Fig. 1(b),(c) and (d).

## 5    Related Work

Early implementations of the fork-join paradigm can be found on Unix and Posix environments, to the state-of-the-art programming models, such as Cilk plus (https://www.cilkplus.org/) and OpenMP (http://www.openmp.org/). An implementation of the framework is available in Java 7. Its focus consists of the optimized usage of all the processor cores, a goal that is accomplished by subdividing recursively the initial task into smaller ones. The process ends when the resulting unit can be processed in asynchronous way. The final result is obtained by applying the *join* construct to merge consistently all the partial results produced by the subtasks.

Fork-join constructs have been implemented by using stochastic formalisms such as Queuing Networks (see [22] and [5]). Again, a fork node spawns the original task in many subtasks, which are directed to a set of queueing stations modeling a set a servers. When all the subtasks have been executed, they are synchronized on the join node. In [5], it is remarked that if the fork-join network consists of more than two queues, then it is not possible to consider a closed-form solution unless the Markov chain underlying the model is considered.

A survey about Process Algebras can be found in [15] and [4]. The authors review how PA matches functional and not functional requirements of distributed systems, stressing also the importance of PA-based stochastic extensions. Particular attention is devoted to the issue of space state explosion occurring when using classical stochastic process algebras, a problem that can be mitigated by using fluid flow approximation.

A comparison between PEPA and General Stochastic Petri Networks (GSPNs) is reviewed in [20]. Provided that both the formalisms are equivalent to a Markov process, the authors present continuities and discontinuities among the formalisms. In first place, the typical concept of *state* (i.e. the number of tokens in a place) is present in GSPN but not explicitly in PEPA (i.e. it can be derived). Furthermore, in terms of modeling abstraction, GSPN provides more flexibility with respect to PEPA. Concerning the fork-join model, the authors show how this paradigm can be easily built in both formalisms. However, this representation of the model is not suited to simulate the behavior of a stage and its task.

A specific language called *DAG-calculus* to model DAGs is presented in [2]. The language includes, among the others, constructs such as fork-join and the more flexible async-finish. The authors perform the translation of fork-join scenarios into Dag calculus by exploiting pure lambda-calculus and considering a construct called *parallel pair*.

The process algebra LOTOS [2] (see, for example, [13]) includes a set of primitives to model fork-join models. Neither dedicated syntax nor additional operators are requested. More specifically, in [19] the authors show how it is possible to describe for-join parallelism natively within PEPA. Describing fork-join mechanisms in a compositional manner is a topic that has been studied by various researchers, for example, in [16]. In addition, the fork-join mechanism is the key operation for workflow systems. In Petri nets, this is well-known by van der Aalst's work on subclasses

---

[2] http://cadp.inria.fr/man/lotos.html.

of Petri nets to describe workflow systems [1]. However, researchers have also considered process algebra for that, for instance in [24]. The language proposed in this paper is designed to allow direct modeling of the most frequent scenario that are found in today's massive parallel computing applications.

## 6    Conclusions

This paper presented a prototype tool, in the form of a specialized modeling language, to support decision processes in administration, resource management and planning of a Big Data environment. The language is specially suited for the case, being semantically close to both Apache Spark typical task structures and PEPA. Future work will include an extension of the framework and a deeper integration within the SIMTHESys modeling framework, by enabling further scaling and analytical approximate or exact solutions of models. Further comparisons with the full set of Process Algebra dialect, to emphasize differences and similarities, is also required.

## References

1. Aalst, W.: The application of Petri Nets to workflow management. J. Circ. Syst. Comput. **8**, 21–66 (1998). https://doi.org/10.1142/S0218126698000043
2. Acar, U.A., Charguéraud, A., Rainey, M., Sieczkowski, F.: Dag-calculus: a calculus for parallel computation. In: Proceedings of the 21st ACM SIGPLAN International Conference on Functional Programming, pp. 18–32. ICFP 2016, ACM, New York, NY, USA (2016). https://doi.org/10.1145/2951913.2951946. http://doi.acm.org/10.1145/2951913.2951946
3. Ardagna, D., et al.: Performance prediction of cloud-based big data applications. In: 9th ACM/SPEC International Conference on Performance Engineering (ICPE 2018) (2018)
4. Baeten, J.C.M.: A brief history of process algebra. Theor. Comput. Sci. **335**(2–3), 131–146 (2005). https://doi.org/10.1016/j.tcs.2004.07.036. http://dx.doi.org/10.1016/j.tcs.2004.07.036
5. Balsamo, S., Harrison, P.G., Marin, A.: Methodological construction of product-form stochastic Petri Nets for performance evaluation. J. Syst. Softw. **85**(7), 1520–1539 (2012). https://doi.org/10.1016/j.jss.2011.11.1042. http://dx.doi.org/10.1016/j.jss.2011.11.1042
6. Barbierato, E., Gribaudo, M., Iacono, M.: Modeling and evaluating the effects of Big Data storage resource allocation in global scale cloud architectures. Int. J. Data Warehouse. Min. **12**(2), 1–20 (2016). https://doi.org/10.4018/IJDWM.2016040101
7. Barbierato, E., Gribaudo, M., Iacono, M.: Exploiting multiformalism models for testing and performance evaluation in SIMTHESys. In: Proceedings of 5th International ICST Conference on Performance Evaluation Methodologies and Tools - VALUETOOLS 2011 (2011)
8. Barbierato, E., Gribaudo, M., Iacono, M.: Modeling apache hive based applications in Big Data architectures. In: Proceedings of the 7th International Conference on Performance Evaluation Methodologies and Tools. ValueTools'13, ICST (Institute for Computer Sciences, Social-Informatics and Telecommunications Engineering), ICST, Brussels, Belgium, Belgium, pp. 30–38 (2013). https://doi.org/10.4108/icst.valuetools.2013.254398

9. Barbierato, E., Gribaudo, M., Iacono, M.: A performance modeling language for Big Data architectures. In: Rekdalsbakken, W., Bye, R.T., Zhang, H. (eds.) European Council for Modeling and Simulation (ECMS), pp. 511–517 (2013)

10. Barbierato, E., Gribaudo, M., Iacono, M.: Performance evaluation of NoSQL Big-Data applications using multi-formalism models. Future Gener. Comput. Syst. **37**, 345–353 (2014). https://doi.org/10.1016/j.future.2013.12.036

11. Barbierato, E., Gribaudo, M., Iacono, M., Marrone, S.: Performability modeling of exceptions-aware systems in multiformalism tools. In: Al-Begain, K., Balsamo, S., Fiems, D., Marin, A. (eds.) Analytical and Stochastic Modeling Techniques and Applications (ASMTA 2011). Lecture Notes in Computer Science, vol. 6751, pp. 257–272. Springer, Berlin, Heidelberg (2011). https://doi.org/10.1007/978-3-642-21713-5_19

12. Bertoli, M., Casale, G., Serazzi, G.: JMT: performance engineering tools for system modeling. ACM SIGMETRICS Perform. Eval. Rev. **36**(4), 10–15 (2009)

13. Bolognesi, T., Brinksma, E.: A Compositional Approach to Performance Modelling. Cambridge University Press, Cambridge (1996)

14. Chiola, G.: A software package for the analysis of generalized stochastic Petri Net models. In: International Workshop on Timed Petri Nets, pp. 136–143 (1985)

15. Gaur, M., Kant, R.: A survey on process algebraic stochastic modelling of large distributed systems for its performance analysis. In: 3rd International Conference on Eco-friendly Computing and Communication Systems (ICECCS), 2014 (2014)

16. Havelund, K., Larsen, K.: The fork calculus. Nord. J. Comput. **1**, 346–363 (1994)

17. Hillston, J.: Process algebras for quantitative analysis. In: Proceedings of the 20th Annual IEEE Symposium on Logic in Computer Science. LICS'05, IEEE Computer Society, Washington, DC, USA, pp. 239–248 (2005). https://doi.org/10.1109/LICS.2005.35. https://doi.org/10.1109/LICS.2005.35

18. Hillston, J.: Introduction to the ISO specification language LOTOS. Comput. Netw. ISDN Syst. **14**(1), 25–59 (1988)

19. Hillston, J., Tribastone, M., Gilmore, S.: Stochastic process algebras: from individuals to populations. Comput. J. **55**(7), 866–881 (2011). https://doi.org/10.1093/comjnl/bxr094. https://doi.org/10.1093/comjnl/bxr094

20. Marsan, M.A., Balbo, G., Conte, G., Donatelli, S., Franceschinis, G.: Modelling with Generalized Stochastic Petri Nets, 1st edn. Wiley, New York, NY, USA (1994)

21. Milner, R.: Communication and Concurrency. Prentice-Hall International, Englewood Cliffs (1989)

22. Osman, R., Harrison, P.G.: Approximating closed fork-join queueing networks using product-form stochastic Petri-nets. J. Syst. Softw. **110**(C), 264–278 (2015). https://doi.org/10.1016/j.jss.2015.08.036. http://dx.doi.org/10.1016/j.jss.2015.08.036

23. Stewart, R., Singer, J.: Comparing Fork/Join and MapReduce. Technical report. HW-MACS-TR-0096, Heriot-Watt University Department of Computer Science (2012). http://www.macs.hw.ac.uk/cs/techreps/docs/files/HW-MACS-TR-0096.pdf

24. Wong, P.Y.H., Gibbons, J.: A process-algebraic approach to workflow specification and refinement. In: Lumpe, M., Vanderperren, W. (eds.) Software Composition, pp. 51–65. Springer, Berlin, Heidelberg (2007). https://doi.org/10.1007/978-3-540-77351-1_5

# Performance Evaluation of Scheduling Policies for the DRCMPSP

Ugur Satic$^{(\boxtimes)}$ ⓘ, Peter Jacko ⓘ, and Christopher Kirkbride ⓘ

Lancaster University, Lancaster LA1 4YW, UK
u.satic@lancaster.ac.uk

**Abstract.** In this study, we consider the dynamic resource-constrained multi-project scheduling problem (DRCMPSP) where projects generate rewards at their completion, completions later than a due date cause tardiness costs and new projects arrive randomly during the ongoing project execution which disturbs the existing project scheduling plan. We model this problem as a discrete Markov decision process and explore the computational limitations of solving the problem by dynamic programming. We run and compare four different solution approaches on small size problems. These solution approaches are: a dynamic programming algorithm to determine a policy that maximises the average profit per unit time net of charges for late project completion, a genetic algorithm which generates a schedule to maximise the total reward of ongoing projects and updates the schedule with each new project arrival, a rule-based algorithm which prioritise processing of tasks with the highest processing durations, and a worst decision algorithm to seek a non-idling policy to minimise the average profit per unit time. Average profits per unit time of generated policies of the solution algorithms are evaluated and compared. The performance of the genetic algorithm is the closest to the optimal policies of the dynamic programming algorithm, but its results are notably suboptimal, up to 67.2%. Alternative scheduling algorithms are close to optimal with low project arrival probability but quickly deteriorate their performance as the probability increases.

**Keywords:** Dynamic programming · Resource constraint · Project scheduling · DRCMPSP

## 1 Introduction

Project management is crucial for many sectors such as engineering services, software development, IT services, construction and R&D, [1,5,8,25]. However it is a very challenging enterprise in that only 40% of projects are completed within their planned time, 46% of projects are completed within their predicted budget and only 36% of projects realise their full benefit [3]. Many factors may bring uncertainty to the project execution plan such as new project arrivals. In this environment, problem size grows and becomes intractable for an exact solution; thus, approximation algorithms are generally preferred. This study applies an

ⓒ Springer Nature Switzerland AG 2020
M. Gribaudo et al. (Eds.): ASMTA 2019, LNCS 12023, pp. 100–114, 2020.
https://doi.org/10.1007/978-3-030-62885-7_8

exact solution method and some approximation methods to project scheduling problems under uncertainty and compare their performances.

"A *project* is a unique, transient endeavour, undertaken to achieve planned objectives, which could be defined in terms of outputs, outcomes or benefits." [2]. A project consists of a collection of *tasks* that are connected via network relationships. A *reward* is released as the outcome of a project completion. A project is completed when all of its tasks are processed and an amount of *resources* (e.g. manpower, equipment) is spent over time to process these tasks. Completion of the project beyond a pre-determined *due date* or to lower standards than agreed may cause penalties and loss of prestige and goodwill which are collectively called the *tardiness cost*.

Determining a task processing order to achieve project goals such as completion in the minimum time or completion within a specific time is called *project scheduling problem* (PSP). The PSP is a vast research area which aims at optimising of project duration, resource allocation and cost evaluation [16]. In this area, the *resource-constrained project scheduling problem* (RCPSP) is one of the most extensively studied research [6]. The common goal of the RCPSP is minimising the completion time and *genetic algorithm* (GA) is the most used solution algorithm for this deterministic problem in the literature [11]. Well-known RCPSP test problems are available at PSPLIB [13], which is an online RCPSP library (http://www.om-db.wi.tum.de/psplib).

Companies usually manage multiple projects simultaneously, and the RCPSP with multiple projects is called *resource-constrained multi-project scheduling problem* (RCMPSP) [1]. The RCMPSP is a generalisation of the RCPSP, which is an NP-hard optimisation problem; thus, RCMPSP and the other generalisation of RCPSPs are also categorised as NP-hard. [7]. Two RCMPSP solution approaches exist: (1) the first approach combines projects in parallel with a dummy start-task and a dummy end-task, then solves the problem as a giant RCPSP, (2) the second approach maintains the multiple projects separately [4]. The general goal of the RCMPSP is minimising the total (for the first approach) or the average (for the second approach) completion time [4]. A RCMPSP library named MPSPLIB is available at "http://www.mpsplib.com/" which contains sets of problems generated by Homberger [10].

The RCPSP and RCMPSP are static, where the data of project arrival times and their type are known before the scheduling begins. However, many companies accept new projects during the processing of ongoing projects [9]. That deviates from the project plan and leads to missed due dates and associated tardiness costs [5]. So, instead of focusing only on completion times, projects are modelled with completion rewards and the objective becomes to maximise the expected profit which is the difference between expected completion rewards and expected tardiness costs. The RCMPSP with uncertain project arrivals and deterministic task durations is called *dynamic* RCMPSP (DRCMPSP). Two main approaches are available for the DRCMPSP; (1) reactive baseline scheduling (e.g. [17]), an approach which generates a baseline schedule and updates it at each project arrival which allows usage of the static RCMPSP methods such as the GA for the DRCMPSP and (2) computation of optimal dynamic policies (e.g. [18]).

In this paper, we consider the DRCMPSP with uncertain project arrivals. We model the problem as an infinite-horizon discrete-time *Markov decision process* (MDP) which is defined by five elements: time horizon, decision state space, action set, transition function and profit function. We generated task processing policies for the DRCMPSP using multiple solution methods. First, we used *dynamic programming value iteration* method to maximise the time-average profit. Second, we used GA to maximise the total completion reward and reactively fixed the schedule distribution for each project arrivals. Third, we used a *rule-based algorithm* (RBA) to generate a policy using the longest task first rule. Finally, we used *worst decision algorithm* (WDA) to generate a non-idling policy which aims to minimise the time-average profit.

We contribute to the literature by (i) developing a DRCMPSP model considering multi-task project types, extending the work of Melchiors et al. [14] who only considered single-task projects, (ii) developing an efficient implementation of the value iteration algorithm in Julia programming language to solve our model with up to 4 project types, (iii) comparing the (exactly) optimal policy of value iteration with the above-mentioned benchmark policies to evaluate the performance gap between solution approaches, and (iv) illustrating that even in simple problems with 2 or 3 project types, the suboptimality gap of benchmark policies commonly used in practice (genetic algorithm and longest-task-first rule) which ignore possibility of new project arrivals is remarkable.

This paper is organized as follows: In Sect. 2, we describe the problem setting, the MDP model. In Sect. 3, we describe the compared algorithms and discuss comparison results in Sect. 4. In Sect. 5, conclusion is presented.

## 2   Methodology

### 2.1   The Problem Setting

The DRCMPSP comprises $J$ project types, and the system capacity for each project type is limited to one. All projects of type $j$ share the same characteristics such as arrival probability $(\lambda_j)$, number of tasks $(I_j)$, task durations $(t_{j,i})$, project network, resource usages $(b_{j,i})$, project due date $(F_j)$, reward $(r_j)$ and tardiness cost $(w_j)$.

A project may arrive to the system at any point during the time unit, which is the duration between two decision epochs. Only one project for each type may arrive per unit time with probability $\lambda_j$ for a project of type $j$. Projects are stored in the system until the end of unit time. Then in the next decision epoch, if the system capacity for newly arrived project type is not full, it will get accepted to the system. Otherwise, it will get rejected.

A type $j$ project consists of $I_j$ tasks. In this problem, tasks are connected sequentially with a successor-predecessor relationship, which defines the project network. Processing task $i$ of project type $j$ requires completion of its predecessor tasks $(\mathcal{M}_{j,i})$ which have an earlier place in the project network. An example project network is shown in Fig. 1.

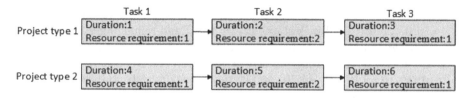

**Fig. 1.** A project network

Processing task $i$ from project type $j$ also requires allocation of $b_{j,i}$ amount of resources during its processing. Only one type of resource is defined in our model and the amount available is represented by $B$. The total number of allocated resources cannot be higher than $B$. The resources are assumed renewable which means they become reusable after completion of a task to which they were assigned. The number of resources which are not allocated for task processing is called free-resources ($B_s^{\text{free}}$). After the completion of a task, its allocated resources return to the free resources.

Task processing is assumed to be non-preemptive; thus, it cannot be paused or cancelled. i.e., once a task has begun processing, it does not leave processing until completed.

Projects are completed when all of their tasks are processed, and a project reward $r_j$ is earned. Projects have a due date ($F_j$) which represents the maximum unit of time which can be spent for project completion to obtain its full reward $r_j$. If the due date is exceeded, the tardiness cost $w_j$ is applied only once, after the project is completed.

## 2.2 Decision State

The decision state ($s$) represents the system information relevant to the decision-making process at each decision epoch [21]. Decision states where the resource limitations are not exceeded and predecessor tasks were completed before their successor tasks are called feasible and the set of all feasible decision states is called as the state space $\mathcal{S}$. Elements of a decision state ($s = \{\boldsymbol{P}_1, \boldsymbol{P}_2 \ldots, \boldsymbol{P}_J\}$) are project states ($\boldsymbol{P}_j$) for all project types. A project state consists of task states ($x_{j,i}$) and the remaining due date state ($d_j$) ($\boldsymbol{P}_j = (x_{j,1}, x_{j,2}, \ldots, x_{j,I_j}, d_j)$).

A task state ($x_{j,i} \in \{-1, 0, 1, 2, \ldots, t_{j,i} - 1\}$) represents the status of a task. If a task is pending for processing, its value is taken as $-1$. If a task is finished, its value is represented by 0. If a task is in processing, its value is the remaining processing time to its completion.

The remaining due date state ($d_j \in \{0, 1, 2, 3, \ldots, F_j\}$) represents the number of remaining time units from the current time epoch to complete the project $j$ without paying any tardiness cost. When a due date is exceeded, its value becomes 0 and it expresses that the tardiness cost will be incurred at the project's completion. A newly accepted project has the highest remaining due date state value which is its due date $F_j$.

When a type $j$ project is completed or there is no type $j$ project in the system $(\boldsymbol{P}_j = (0,0,\ldots,0,0))$, all task states $(x_{j,i} = 0, \forall i)$ and remaining due date state $(d_j = 0)$ of project type $j$ are represented by 0.

When a new type $j$ project arrives $(\boldsymbol{P}_j = (-1,-1,\ldots,-1,F_j))$, all its task states are set to $-1$ $(x_{j,i} = -1, \forall i)$ and its remaining due date state is represented by its due date $F_j$ $(d_j = F_j)$.

An example state matrix with two projects and three tasks is shown in Table 1. Here, rows of the matrix represent each project type $j$. The columns represent the task numbers but the last column of the matrix represents the due date state $(d_j)$.

**Table 1.** State matrix

|  |  | Remaining duration of task i | | | Remaining due date |
|---|---|---|---|---|---|
|  |  | 1 | 2 | 3 | d |
| Project j | 1 | $x_{1,1}$ | $x_{1,2}$ | $x_{1,3}$ | $d_1$ |
|  | 2 | $x_{2,1}$ | $x_{2,2}$ | $x_{2,3}$ | $d_2$ |

A decision state determines its free resources $(B_s^{\text{free}})$ which is used as a constraint for the available decisions. Free resources are the remaining resources available after the resource allocation to ongoing tasks has been accounted for:

$$B_s^{\text{free}} = B - \sum_{j=1}^{J} \sum_{i=1}^{I_j} b_{i,j} \mathcal{I}\{x_{i,j} > 0\} \tag{1}$$

Here, $B$ is the total amount of resource, $b_{i,j}$ is the resource amount allocated for processing of task $i$ from type $j$ project, $\mathcal{I}\{.\}$ is an indicator function that takes the value 1 if the condition in parentheses is true and takes the value 0 otherwise.

## 2.3 Action Representation

The decisions available in a given decision state $s$ is called an action $a$. At a decision epoch, the decision maker selects an action $a$ which starts the processing of the selected pending tasks. An example action matrix with two projects and three tasks is shown in Table 2. If the decision includes processing a pending task $i$ of a type $j$ project $(x_{i,j} = -1)$, the corresponding action element $a_{j,i}$ will take the value of 1 in the action matrix. Otherwise, $a_{j,i}$ will be 0. The task processing decision can be only taken if there are enough free resources to allocate $(\sum_{j=1}^{J} \sum_{i=1}^{I_j} b_{i,j} \mathcal{I}\{a_{i,j} = 1\} \le B_s^{\text{free}})$ and any predecessor tasks $(\mathcal{M}_{j,i})$ of task $i$ are completed $(\sum_{m \in \mathcal{M}_{j,i}} x_{j,m} = 0)$. Thus, an action must satisfy both of these conditions. All the actions which meet both the resource and predecessor limitations, are called feasible and set of all feasible actions for a decision state $s$ creates the action set $(A(s) = \{0, a', a'', \dots\})$.

**Table 2.** Action matrix

|  |  | Task i |  |  |
|---|---|---|---|---|
|  |  | 1 | 2 | 3 |
| Project j | 1 | $a_{1,1}$ | $a_{1,2}$ | $a_{1,3}$ |
|  | 2 | $a_{2,1}$ | $a_{2,2}$ | $a_{2,3}$ |

The action, where all action elements are zero "do not initiate any task" $(0 = (0, 0, \dots, 0))$ and it is always a member of the action set $0 \in A(s)$. $a'$ and $a''$ represent alternative feasible actions. The number of alternative actions in an action set depends on the number of free resources $(B_s^{\text{free}})$, the unprocessed tasks (with $x_{i,j} = -1$) and the tasks with completed predecessor tasks (with $\sum_{m \in \mathcal{M}_{j,i}} x_{j,m} = 0$).

## 2.4 Transition Function

The transition function describes how the system evolves from one state to another as a result of decisions and information [19]. The period between two consecutive decision states is the time unit. During the transition period; the ongoing tasks are processed for one time unit, some tasks are completed and new projects may arrive according to arrival probabilities $\lambda_j$. The project arrival probability is considered when a project is to be completed before the next decision epoch (i.e., the system capacity for type-$j$ project will become available). The transition function is defined in Eq. 2.

$$P(s'|s,a) = \prod_{j=1}^{J} \prod_{i=1}^{I_j} P(x'_{j,i}|x_{j,i}) \tag{2}$$

$$P(x'_{j,i}|x_{j,i}) = \begin{cases} \lambda_j, & \text{for } 0 \le x_{j,i} \le 1, x'_{j,i} = -1, i = I_j \\ \lambda_j, & \text{for } x_{j,i} = -1, a_{j,i} = 1, x'_{j,i} = -1, i = I_j \\ 1 - \lambda_j, & \text{for } 0 \le x_{j,i} \le 1, x'_{j,i} = 0, i = I_j \\ 1 - \lambda_j, & \text{for } x_{j,i} = -1, a_{j,i} = 1, x'_{j,i} = 0, i = I_j \\ 1, & \text{for } x_{j,i} \ge 2, x'_{j,i} = x_{j,i} - 1, i = I_j \\ 1, & \text{for } x_{j,i} = -1, a_{j,i} = 0, x'_{j,i} = -1, i = I_j \\ 1, & \text{for } x_{j,i} \ge -1, x'_{j,i} \ge -1, i < I_j \end{cases}$$

In Fig. 2 an example transition process has been shown. The transition probability of the first alternative future decision state, where the last task of type $j$ project is finished and a new type $j$ project arrived, is $P(s''|s, a) = \lambda_j$. The transition probability of the second alternative future decision state, where the last task type $j$ project is finished and no project arrived, is $P(s'|s, a) = (1 - \lambda_j)$.

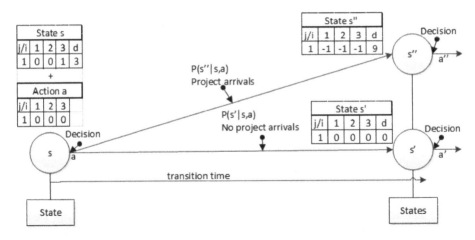

**Fig. 2.** A state transition diagram (for a $j = 1$ type project with 3 tasks ($i = 1, 2, 3$) whose due date is $F_j = 9$ and the selected action means do not initialise any task.)

## 2.5   Profit Representation

The profit function ($R_{s,a}$) is the sum of rewards ($r_j$) of completed projects in the period between current and next decision epoch minus the tardiness cost of late completions which depend on the remaining due dates.

$$R_{s,a} = \sum_{j=1}^{J} r_j \mathbb{E}\Big[\mathcal{I}\{x_{j,I} = 1 \vee (x_{j,I} = -1 \wedge a_{j,I} = 1 \wedge t_{j,I} = 1)\}\Big]$$
$$- \sum_{j=1}^{J} w_j \mathbb{E}\Big[\mathcal{I}\{x_{j,I} = 1 \vee (x_{j,I} = -1 \wedge a_{j,I} = 1 \wedge t_{j,I} = 1) \wedge d_j = 0\}\Big]$$

$$(3)$$

Here, the first indicator is for project completion and takes the value 1 if a project completes and is 0 otherwise. The second indicator is for late project completion. It takes the value 1 if a project's due date has already passed (i.e., the project's remaining dues date $d_j = 0$) and is 0 otherwise. Recall that, in decision state $s$, $x_{j,I}$ represents the remaining processing time of the final task of a type $j$ project and $a_{j,I}$ its the action element under action $a$. $t_{j,I}$ is duration of task $i$ of project type $j$.

## 2.6   Goal Function

The goal of the DRCMPSP is to find the policy $\pi$ that maximises the long-term average profit per unit time.

$$g^* = \max_{\pi \in \Pi} \lim_{T \to \infty} \frac{1}{T} \sum_{t=1}^{T} \mathbb{E}^{\pi}[R_{s(t),a(t)}]$$

$$(4)$$

Here, $t$ is the time epoch. $R_{s(t),a(t)}$ is the profit function dependent of time epoch $t$. $\pi$ is a policy from the set of all feasible non-anticipating policies ($\Pi$) presenting the action set $A(s)$. A feasible policy is a sequence of action which considers both the resource limitation and project network.

## 2.7   Solution by Dynamic Programming

Dynamic Programming is a collection of algorithms which calculates optimal policies from the MDP model of the solution environment [23]. In this research we used Dynamic Programming Value Iteration. Value Iteration calculates a sequence of value functions [24]. The value function approximates the cumulative reward minus the tardiness cost. The per-period change in the value function approximates the maximum long-term average profit. The process steps of the algorithm are below;

For each state $\forall s \in \mathcal{S}, V^{old}(s) = 0$
Do
$\quad$ For each state $\forall s \in \mathcal{S}$
$\quad\quad V(s) = \max_{a \in A}[R_{s,a} + \sum_{s' \in \mathcal{S}} p(s'|s,a)V^{old}(s')]$
$\quad$ End For
$\quad W_{max} = \max_{s \in \mathcal{S}}[V(s) - V^{old}(s)]$
$\quad W_{min} = \min_{s \in \mathcal{S}}[V(s) - V^{old}(s)]$
$\quad \Delta = W_{max} - W_{min}$
$\quad$ Update for $\forall s \in \mathcal{S}, V^{old}(s) = V(s)$
While $\Delta > \beta \times W_{min}$

Here, $V$ represents the value function of a decision state $s$. $R_{s,a}$ is profit function as explained in Subsect. 2.5. $p(s'|s,a)$ is the state transition probability. $s'$ stands for the future decision state of $s$. $V^{old}(s')$ is the value of $s'$ from next decision epoch. $\beta$ is pre-specified tolerance number (0.000001). $W_{min}$ and $W_{max}$ are respectively minimum and maximum value changes between two iterations. $\Delta$ is the difference between the minimum and the maximum value changes. $\mathcal{S}$ is the state space which is defined at Subsect. 2.2. These processes are repeated until the stopping criteria is met.

# 3   Results and Comparisons

We used two heuristic algorithms with reactive scheduling and one worst decision algorithm to compare their performance to optimal. A reactive scheduling method generates decisions within a deterministic approach without considering the future uncertainties [17]. Then, it iteratively fixes its first schedule according to random changes and makes the schedules feasible again [20]. We used a genetic algorithm and a priority rule algorithm; note that both are based on the reactive scheduling method. Both the optimal DP and the worst decision algorithm are proactive scheduling methods.

## 3.1  Genetic Algorithm

The discrete-time MDP is considered as a reactive scheduling system by generating a new baseline schedule for each decision state. The baseline schedules are generated by a genetic algorithm (GA) which seeks to maximise the profit and minimising the total completion time. We adapted GA from Satic [22]. The GA is one of the search algorithms which searches for the global optimum on the solution space by improving the search samples at each iteration [15]. The GA uses bio-inspired operators (e.g. Elitist selection, Crossover and Mutation) to develop the population, which is a solution set, in each iteration.

For each decision state, random numbers are assigned to unprocessed tasks, and this assignment is stored as an individual of the population. Individuals are created until the population number (here, one hundred) is reached. The random numbers represent task processing priorities and this method called as the random key representation. The random keys are converted to a schedule using the serial scheduling scheme as Kolisch and Hartmann [12] described. Then the population is ordered according to their total profit and total completion time.

The first population is iterated one hundred times using the genetic operators. The best ten percent of the population is transferred to the next population without any change, and the rest of the next population is created with the crossover operator. The crossover operator, firstly, selects two individuals from the previous population, then, copies some random keys from the first individual, after that, copies the rest from another individual, and finally, creates a new individual. The new individual is mutated with a fifty per cent probability before joining to the next population. The mutation operator randomly selects an unprocessed task and re-assigns its random number. When the new population reaches one hundred individuals, the random keys are converted to schedules with the serial scheduling scheme and the population is ordered again according to their total profit and total completion time. After the one-hundredth generation is created; the best schedule is selected as the baseline schedule. Then the baseline schedule is converted to action.

## 3.2  Priority Rule (Longest Task First)

An alternative policy is created with a priority based heuristic algorithm. The algorithm uses a single-pass priority rule called the longest task first rule. Single-pass rules generate only one action for the given state. The rule based algorithm (RBA) prioritises the tasks with the longer processing times and if two tasks have the same duration, the smallest numbered project type, e.g., project type 1 is prioritised over type 2 or type 3. For each decision state, the algorithm generates a baseline schedule using the priority rule and the serial scheduling scheme. Then, the baseline schedule is converted to an action.

### 3.3    Worst Decision Algorithm

A mix of value iteration and priority rule methods are used as the worst decision algorithm (WDP) which seeks a policy ($\pi'$) to get the minimum profit per unit time.

$$g' = \min_{\pi' \in \Pi'} \lim_{T \to \infty} \frac{1}{T} \sum_{t=1}^{T} \mathbb{E}^{\pi'}[R_{s(t),a(t)}]. \tag{5}$$

Here, $\pi'$ is a policy from the set of all feasible non-anticipating active policies ($\Pi'$) which does not include the "to do not activate any task" ($\mathbf{0}$) actions unless it is the only possible action in the action set ($|\boldsymbol{A}(s)| = 1$). Since the reward and tardiness costs are modelled to be received after project completions, a minimum profit algorithm without the priority rule ($|\boldsymbol{A}(s)| \neq 1 \Rightarrow \mathbf{0} \notin \pi'$) delays project completions infinitely to halt rewards.

## 4    Computational Results

### 4.1    Experimental Setup

In this section, we explore the limits of DP on the DRCMPSP, and compare its performance with the two heuristic reactive baseline scheduling algorithms and the worst decision algorithm. The DP and the compared algorithms are coded in JuliaPro 1.0.1.1. All tests are performed on a desktop computer with Intel i5-6500T CPU with 2.50 GHZ clock speed and 32 GB of RAM.

We generate four DRCMPSPs (see Table 3). For each project in the experiment, a project's tasks are performed in sequential numerical order, i.e., a project starts with task one which is a predecessor of task two which is a predecessor of task three. See Fig. 1. The problems vary by number of projects, number of tasks, resource usage, different reward-tardiness cost settings and length of due date. We call the difference between a project's due date and the sum of tasks durations as slack time. This value also varies for each project in the problems. The total resource capacity is taken $B = 3$ for all problems.

The first problem has two project types, and each type has two tasks. Project type two has a higher completion reward and higher tardiness cost with a shorter slack time. That means while project type two contributes higher reward opportunities, its late completion is less rewarding compared to the late completion of the project type one.

The second problem has two project types, and each type has three tasks. The project type one is as twice as profitable. However, the slack time of project type one is shorter, so its due date may easily be exceeded leading to tardiness cost.

The third problem has three projects types, and each type has two tasks. In this problem, resource capacity allows parallel processing for only up to two projects increasing the chance of tardiness costs from one project. Only project type one can be processed with other types.

**Table 3.** Problem parameters

| 2 projects and 2 tasks problem | | | | | | |
|---|---|---|---|---|---|---|
| Project no | Reward | Tardiness cost | Due date | Task no | Task duration | Resource usage |
| 1 | 3 | 1 | 8 | 1 | 2 | 2 |
| | | | | 2 | 2 | 2 |
| 2 | 10 | 9 | 5 | 1 | 3 | 1 |
| | | | | 2 | 1 | 3 |

| 2 projects and 3 tasks problem | | | | | | |
|---|---|---|---|---|---|---|
| Project no | Reward | Tardiness cost | Due date | Task no | Task duration | Resource usage |
| 1 | 12 | 8 | 10 | 1 | 1 | 1 |
| | | | | 2 | 2 | 2 |
| | | | | 3 | 5 | 1 |
| 2 | 6 | 5 | 15 | 1 | 4 | 1 |
| | | | | 2 | 3 | 2 |
| | | | | 3 | 4 | 1 |

| 3 projects and 2 tasks problem | | | | | | |
|---|---|---|---|---|---|---|
| Project no | Reward | Tardiness cost | Due date | Task no | Task duration | Resource usage |
| 1 | 8 | 5 | 10 | 1 | 5 | 1 |
| | | | | 2 | 2 | 1 |
| 2 | 5 | 3 | 8 | 1 | 1 | 2 |
| | | | | 2 | 3 | 1 |
| 3 | 20 | 19 | 10 | 1 | 2 | 3 |
| | | | | 2 | 7 | 2 |

| 4 projects and 2 tasks problem | | | | | | |
|---|---|---|---|---|---|---|
| Project no | Reward | Tardiness cost | Due date | Task no | Task duration | Resource usage |
| 1 | 18 | 3 | 4 | 1 | 5 | 2 |
| | | | | 2 | 1 | 1 |
| 2 | 27 | 4 | 5 | 1 | 4 | 2 |
| | | | | 2 | 2 | 1 |
| 3 | 18 | 5 | 6 | 1 | 3 | 2 |
| | | | | 2 | 3 | 1 |
| 4 | 18 | 6 | 7 | 1 | 2 | 2 |
| | | | | 2 | 4 | 1 |

*Resource capacities = 3

The fourth problem has four projects types, and each type has two tasks. The slack times of project types one and two are negative, and project type three's slack times is zero and project type four's slack time is one. Thus most of the projects will be completed later than their planned due date, and the tardiness payment will be inevitable.

We test each problem consecutively from 1% to 90% project arrival probabilities, increment by 10%. 0% and 100% arrival probabilities are not used in this comparison, because 0% arrival probability makes the problem static and 100% arrival probability causes a non-ergodic MDP, e.g., the empty state where no project has arrived cannot be reachable again from any states.

**Table 4.** Comparison of the time-average profit deviations from the optimal results of DP (how much percent lower than optimal results of DP)

| | Project arrival probability | | | | | | | | | |
|---|---|---|---|---|---|---|---|---|---|---|
| | 1% | 10% | 20% | 30% | 40% | 50% | 60% | 70% | 80% | 90% |
| | 2 projects and 2 tasks problem | | | | | | | | | |
| GA | 0.7% | 6.5% | 11.7% | 15.4% | 18.1% | 20.0% | 21.2% | 21.4% | 20.4% | 17.0% |
| RBA | 2.1% | 19.9% | 35.2% | 46.1% | 53.7% | 59.3% | 63.7% | 67.3% | 70.4% | 72.7% |
| WDP | 2.8% | 25.6% | 43.8% | 55.4% | 62.7% | 67.3% | 70.2% | 72.1% | 73.5% | 75.5% |
| | 2 projects and 3 tasks problem | | | | | | | | | |
| GA | 0.1% | 4.9% | 13.0% | 22.0% | 31.1% | 39.1% | 45.6% | 51.6% | 58.1% | 67.2% |
| RBA | 1.5% | 15.3% | 25.1% | 30.1% | 32.3% | 32.6% | 31.1% | 28.2% | 23.5% | 15.4% |
| WDP | 4.1% | 34.3% | 49.9% | 59.0% | 66.4% | 72.6% | 77.1% | 80.2% | 82.2% | 83.3% |
| | 3 projects and 2 tasks problem | | | | | | | | | |
| GA | 0.1% | 4.9% | 13.0% | 22.0% | 31.1% | 39.1% | 45.6% | 51.6% | 58.1% | 67.2% |
| RBA | 1.5% | 15.3% | 25.1% | 30.1% | 32.3% | 32.6% | 31.1% | 28.2% | 23.5% | 15.4% |
| WDP | 4.1% | 34.3% | 49.9% | 59.0% | 66.4% | 72.6% | 77.1% | 80.2% | 82.2% | 83.3% |
| | 4 projects and 2 tasks problem | | | | | | | | | |
| GA | 0.0% | 1.2% | 2.9% | 5.8% | 6.9% | 6.8% | 8.0% | 11.5% | 15.4% | 19.0% |
| RBA | 0.4% | 6.6% | 14.6% | 21.4% | 25.1% | 26.8% | 28.7% | 31.4% | 33.9% | 36.1% |
| WDP | 1.4% | 21.3% | 37.8% | 46.2% | 50.5% | 52.8% | 54.8% | 57.3% | 59.4% | 61.5% [a] |

[a] approximate

## 4.2   Discussion

DP suffers from "the curse of dimensionality" which means, here, the number of states grows exponentially with the number of tasks in a project, the number of project types, task durations and due dates, and the large state space becomes computationally intractable [23]. The model uses the state space as defined in Subsect. 2.2. In our experiment, a state space for more than five project types with two tasks each becomes computationally intractable. Thus the considered problems are limited to four projects and two tasks.

The results shown in Table 4 illustrate that the GA produces almost optimal solutions in 1% arrival probability and close to optimal solutions with other low arrival probabilities. The GA's results are generally closer to optimum compared to RBA for the majority of the considered problems and their task duration variations. The GA's results were from 0.003% to 67.2% lower than the optimum results but never exactly the same.

The RBA's results are between the GA and the WDP for most of the test problem. The RBA's results were from 0.4% to 72.7% lower than the optimum results. In three projects with two tasks problem, the RBA produced better results than the GA at higher arrival probabilities. However, in most of the cases, its results were closer to the WDP than the optimum since the used priority rule is not designed for reward maximising.

Since the GA and the RBA are reactive baseline scheduling algorithms, they generate decisions without considering the new project arrivals. Thus we may accept that the result of a reactive baseline scheduling algorithm deteriorates compared to the optimum as problem deviates from the static assumption i.e. no project arrivals. However, some anomalies were observed for very high arrival probabilities. These anomalies occur since the tardiness cost is only paid once when a project is completed. In the current model, high arrival probabilities lead to postponing some projects infinitely. Thus, they stay in the system without causing a tardiness cost while the other projects continue processing without causing much tardiness cost.

## 5    Conclusion

In this paper, we studied the resource-constrained multi-project scheduling problem with uncertain project arrivals. We modelled the problem as an infinite-horizon discrete-time MDP. New project arrivals happen during the time unit. We used DP value iteration to maximise the long-term average profit per unit time. We tested the limits of the DP on the DRCMPSP and generated four test problems. We used two heuristic reactive baseline scheduling methods and a worst-decision DP on the same problems and compared their results with exact results of the DP. We used GA and RBA as heuristic reactive baseline scheduling methods.

According to our findings, GA produced closer to optimal results than the simpler heuristic RBA. Since reactive baseline scheduling does not consider the random changes before they occurred, the GA's and the RBA's results are closer to optimal at low arrival probabilities, and diverge from optimum at the high arrival probabilities.

In this work, we have seen that DP suffers from the curse of dimensionality even for the small size problems and reactive baseline scheduling methods do not produce close to optimum results at the high arrival probabilities. Therefore, as a future research topic, we suggest to use a technique which will not (or less) suffer from the curse of dimensionality but will consider the new project arrivals during the decision phase.

## References

1. Adhau, S., Mittal, M.L., Mittal, A.: A multi-agent system for distributed multi-project scheduling: an auction-based negotiation approach. Eng. Appl. Artif. Intell. **25**(8), 1738–1751 (2012). https://doi.org/10.1016/j.engappai.2011.12.003
2. APM: APM body of knowledge. Association for Project Management, 6th edn. (2012)
3. APM: The state of project management annual survey 2018 (2018). http://www.wellingtone.co.uk/wp-content/uploads/2018/05/The-State-of-Project-Management-Survey-2018-FINAL.pdf
4. Browning, T.R., Yassine, A.A.: Resource-constrained multi-project scheduling: priority rule performance revisited. Int. J. Prod. Econ. **126**(2), 212–228 (2010). https://doi.org/10.1016/j.ijpe.2010.03.009

5. Capa, C., Kilic, K.: Proactive project scheduling with a bi-objective genetic algorithm in an R&D department. In: 2015 International Conference on Industrial Engineering and Operations Management (IEOM). vol. 1, pp. 1–6 (2015). https://doi.org/10.1109/IEOM.2015.7093733

6. Creemers, S.: Minimizing the expected makespan of a project with stochastic activity durations under resource constraints. J. Sched. **18**(3), 263–273 (2015). https://doi.org/10.1007/s1095

7. Gonçalves, J.F., Mendes, J.J., Resende, M.G.: A genetic algorithm for the resource constrained multi-project scheduling problem. Eur. J. Oper. Res. **189**(3), 1171–1190 (2008). https://doi.org/10.1016/j.ejor.2006.06.074

8. Grey, J.R.: Buffer techniques for stochastic resource constrained project scheduling with stochastic task insertions problems. Ph.D. thesis, University of Central Florida (2007). http://purl.fcla.edu/fcla/etd/CFE0001584

9. Herbots, J., Herroelen, W., Leus, R.: Dynamic order acceptance and capacity planning on a single bottleneck resource. Nav. Res. Logist. (NRL) **54**(8), 874–889 (2007). https://doi.org/10.1002/nav.20259

10. Homberger, J.: A $(\mu, \lambda)$-coordination mechanism for agent-based multi-project scheduling. OR Spectr. **34**(1), 107–132 (2012)

11. Karam, A., Lazarova-Molnar, S.: Recent trends in solving the deterministic resource constrained project scheduling problem. In: 9th International Conference on Innovations in Information Technology (IIT), pp. 124–129 (2013). https://doi.org/10.1109/Innovations.2013.6544405

12. Kolisch, R., Hartmann, S.: Heuristic algorithms for the resource-constrained project scheduling problem classification and computational analysis. In: Węglarz, J. (ed.) Project Scheduling, pp. 147–178. Springer, Boston (1999). https://doi.org/10.1007/978-1-4615-5533-9_7

13. Kolisch, R., Sprecher, A.: PSPLIB-a project scheduling problem library: OR software-ORSEP operations research software exchange program. Eur. J. Oper. Res. **96**(1), 205–216 (1997). https://doi.org/10.1016/S0377-2217(96)00170-1

14. Melchiors, P., Leus, R., Creemers, S., Kolisch, R.: Dynamic order acceptance and capacity planning in a stochastic multi-project environment with a bottleneck resource. Int. J. Prod. Res. **56**(1–2), 459–475 (2018). https://doi.org/10.1080/00207543.2018.1431417

15. Mori, M., Tseng, C.C.: A genetic algorithm for multi-mode resource constrained project scheduling problem. Eur. J. Oper. Res. **100**(1), 134–141 (1997). https://doi.org/10.1016/S0377-2217(96)00180-4

16. Ortiz-Pimiento, N.R., Diaz-Serna, F.J.: The project scheduling problem with non-deterministic activities duration: a literature review. J. Ind. Eng. Manag. (JIEM) **11**(1), 116–134 (2018). https://doi.org/10.3926/jiem.2492

17. Pamay, M.B., Bülbül, K., Ulusoy, G.: Dynamic resource constrained multi-project scheduling problem with weighted earliness/tardiness costs. In: Pulat, P.S., Sarin, S.C., Uzsoy, R. (eds.) Essays in Production, Project Planning and Scheduling. ISORMS, vol. 200, pp. 219–247. Springer, Boston (2014). https://doi.org/10.1007/978-1-4614-9056-2_10

18. Parizi, M.S., Gocgun, Y., Ghate, A.: Approximate policy iteration for dynamic resource-constrained project scheduling. Oper. Res. Lett. **45**(5), 442–447 (2017). https://doi.org/10.1016/j.orl.2017.06.002

19. Powell, W.B.: Approximate Dynamic Programming: Solving the Curses of Dimensionality. Wiley Series in Probability and Statistics, vol. 842. Wiley (2011). https://doi.org/10.1002/9781118029176

20. Rostami, S., Creemers, S., Leus, R.: New strategies for stochastic resource-constrained project scheduling. J. Sched. **21**(3), 349–365 (2018). https://doi.org/10.1007/s1095
21. Sammut, C., Webb, G.I. (eds.): Encyclopedia of Machine Learning. Decision Epoch. Springer, Boston (2010). https://doi.org/10.1007/978-0-387-30164-8_198
22. Satıç, U.: Çok kaynak kısıtlı projelerin sezgisel yöntemlerle çizelgelenmesi. Yildiz Technical University (2014). https://tez.yok.gov.tr/UlusalTezMerkezi/TezGoster?key=gyLHMouPes-CvnhRcjQsKQIo1AYi4WhfoZxbAaK4INn4yvJXWUop3HXf7wZV_sh4
23. Sutton, R.S., Barto, A.G.: Reinforcement Learning: An Introduction. Adaptive Computation and Machine Learning, 2nd edn. MIT Press, Cambridge (2018)
24. Tijms, H.C.: Stochastic Models: An Algorithmic Approach. John Wiley & Sons, Hoboken (1994)
25. Wang, X., Chen, Q., Mao, N., Chen, X., Li, Z.: Proactive approach for stochastic RCMPSP based on multi-priority rule combinations. Int. J. Prod. Res. **53**(4), 1098–1110 (2015). https://doi.org/10.1080/00207543.2014.946570

# An Algorithm for Improved Proportional-Fair Utility for Vehicular Users

Thi Thuy Nga Nguyen[1,2(✉)], Olivier Brun[2], and Balakrishna J. Prabhu[2]

[1] Continental Digital Service in France, Toulouse, France
nttnga@math.ac.vn

[2] LAAS-CNRS, Université de Toulouse, CNRS, Toulouse, France
{brun,Balakrishna.Prabhu}@laas.fr

**Abstract.** The Proportional Fair (PF) scheduler currently implemented in cellular networks is optimal when the channel conditions are stationary. Using measurements, a recent work shows that the conditions for moving cars can be non stationary and vary along the route. Based on these observations, the authors of [8] devise an algorithm called $(PF)^2S$ that exploits Signal-to-Noise Ratio (SNR) maps and rate predictions to improve the utility over the standard PF algorithm. We propose an algorithm which gives a better prediction of the future rate allocation and has a better utility compared to both the PF and $(PF)^2S$ algorithms. The proposed algorithm employs projected gradient on a relaxed version of the problem to predict the future allocations. Simulation results show that non negligible gains in utility over $(PF)^2S$ can be achieved by this algorithm.

**Keywords:** Scheduling · Proportional fairness · Projected gradient

## 1 Introduction

Connected vehicles are expected to have more stringent QoS requirements than that of typical mobile users of today. These will include a higher data rate, reduced latency, and very low outage. In order to meet these requirements, 5G and future cellular technologies will operate in a range of spectrum with sufficiently large bandwidth to accommodate the requests. While increasing the available wireless resources is indeed necessary, it is also important to design resource sharing algorithms that can maximize the utilization of these resources.

The scheduling algorithm in 3G systems is based on the Proportional Fair (PF) scheduler which is obtained as the solution of a utility maximization problem [6]. It was designed with the objective of being fair to users with different channel conditions. Indeed, due to the various wireless effects such as shadowing and fading, data rates of mobile users can vary widely within a zone. By giving

This work was partially funded by a contract with Continental Digital Services France.

the channel to the user with the highest current data rate, the mobile operator will be unfair to users who find themselves in unfavorable channel conditions over long periods of time. The PF scheduler alleviates this problem by allocating the channel to the user with the highest ratio of current data rate to the previous throughput[1] (we will call this ratio the index of a user). Thus, users with comparatively low throughput are assigned a higher priority even when they are in worse channel conditions. The performance and design of the proportional fair scheduler has been widely investigated for wireless networks [2,11].

Proportional Fair scheduling algorithms for wireless networks have been widely investigated in various settings [3,4,12,13]. Most of the literature is based on the assumption that users experience stationary channel conditions. This was partly motivated by the fact that a simple index-based allocation algorithm had been shown to be optimal for stationary channels [7]. This assumptions is not necessarily true for vehicular traffic moving along a given path, as was shown in [8] using SNR maps obtained by measurements. Indeed, as a car move along a road, the SNR improves as it moves closer to a base station and then worsens as it moves away. This implies that SNR is not stationary since its mean varies with time. The long sojourn time assumption is also not realistic for mobile users, since vehicles pass the coverage range of the base station in more or less than one minute. Moreover, if the trajectory of a car is known, one can obtain good statistical predictions on the SNR that will be experienced by the car. Knowing the future channel conditions, one can hope to design scheduling algorithms that can obtain a higher utility compared to the PF algorithm. This improvement in utility was not possible for stationary channel conditions as knowing the current position in the trajectory did not bring any new information on the future data rates.

Several variations of opportunistic scheduling algorithms have been studied recently using future information [1,8]. In [1], the authors use a different kind of proportional fair objective function to ours, and they use future information by looking at channel state of users in a few small time-slots. Different from their approach, we do not look at the predicted channel state in few time slots which may be different between users and difficult to predict correctly due to fast fading. Instead, we base our allocation on average rate the user will experience during the time interval this user stays inside the coverage range of the base station. The average rate in the future is easier to estimate, has a lower error in prediction, and gives useful information for how much data the user can receive. In [8], the authors investigate the same objective function to ours, and propose an improved scheduling algorithm, called $(PF)^2S$ (to be explained later), based on prediction of future rates. In brief, in every time slot the channel is allocated to the user with highest index which is the ratio of the current data rate and the total throughput. The main difference with the PF algorithm is that the total throughput includes the future predicted throughput whereas in

---

[1] The throughput is different from the data rate. While the latter is potential rate at which a user can be served, the former can be smaller since in some slots a user may not be served due to the presence of other users.

the PF algorithm only the past throughput was used. It was shown that this new index led to improved utility compared to the PF algorithm in non-stationary environments.

The $(PF)^2S$ algorithm would have been optimal if the future throughput could be computed optimally. For this, one requires the exact knowledge of all the channel conditions in the future. Since this information is not available, $(PS)^2S$ uses allocations like round-robin (and some other heuristics) for the computation of the future throughput. That is, it is assumed that cars would be served in a round-robin fashion in future time-slots. Clearly, this may not be true since cars are actually served using the index policy of $(PF)^2S$. Nevertheless, it was shown that even without the knowledge of the future throughput, improvements in the utility were made compared PF.

### 1.1 Contributions

We present a heuristic algorithm for non stationary channels that improves the total utility of users compared to the PF and the $(PF)^2S$ algorithms. The original utility maximization problem being computationally complex, we employ three techniques to obtain a lower complexity heuristic: *(i)* we relax the integer constraints of the original problem; *(ii)*, we shorten the time horizon over which the problem is solved; and *(iii)* we compute the solution over macroscopic time slots instead of microscopic ones.

The relaxation turns the problem into a convex one and allows for its efficient resolution. Shortening of the time horizon and solving over macroscopic slots reduces the number of variables in the problem and decreases the computation time.

Simulation results for a single base station scenario show that improvements are achieved in different scenarios with vehicles moving at either equal or different speeds.

### 1.2 Organisation

In Sect. 2, we state the assumptions, define the objective function, and give some background on PF and $(PF)^2S$ algorithms. In Sect. 3, we present the our heuristic for improving the utility based on estimations of future average data rate. Section 4 contains the numerical results for scenarios with homogeneous as well as heterogeneous vehicles. Finally, we end the paper in Sect. 5 with a few open problems.

## 2 Problem Formulation

We consider a single base station (BS) that covers a linear stretch of road of length $L$ (for example $L = 1$ km) along which vehicles move in one direction. The users enter the coverage range of that BS at left edge, move at different velocities, and leave when they arrive at the right edge. Every $\delta = 2$ ms the

BS[2] has to decide which user to serve. Let $v$ be the velocity of the users. The coverage range can be chopped into $N$ small spatial slots with each spatial slot corresponding to distance moved in $\delta$. (See Fig. 1). We also define a big time slot of length $\Delta = 1\,\mathrm{s}$ in which a new car enters the coverage range from the left with some probability.

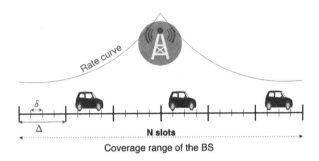

**Fig. 1.** Coverage model for moving vehicles

We shall assume that the data rate received by a user in spatial-slot $s$ depends on the distance between the BS and $s$. The data rate depends upon the SNR which itself can vary along the road. In our numerical experiments, we assume that the data rate decays exponentially as shown in Fig. 1. The scheduling algorithm we propose does not require this assumption to work.

Denote by $T$ (T should be large compared to $\delta$) the time horizon over which the scheduling decisions are made. Let $K$ be number of total users who pass by the BS in $T$. Let $r_{i,j}$ be the feasible data rate of user $i$ in time-slot $j$, and let $\alpha_{i,j} \in \{0,1\}$ denote whether user $i$ is served in slot $j$ or not. Denote by $\alpha = \{\alpha_{i,j}\}_{i,j}$ allocation matrix, and $r = \{r_{i,j}\}_{i,j}$ rate matrix. Our objective is to achieve the proportional-fairness between users, which is described by the following optimization problem (see, e.g., [2,8]):

$$(I)\begin{cases} \max O(\alpha) = \sum_{i=1}^{K} \log\left(\sum_{j=1}^{T} \alpha_{ij} r_{ij}\right) \\ \text{subject to } \sum_{i=1}^{K} \alpha_{ij} = 1,\, j = 1,\dots,T \\ \text{and } \alpha_{ij} \in \{0,1\} \end{cases}$$

Proportional fairness is a resource allocation algorithm which is a special case of the general framework of utility maximization problem [10]. It was first introduced for wired in [6] and has been applied in various wireless settings [2,11]. In brief, each user stays in the system for a certain duration during which it receives certain throughput which depends on the allocation decisions of the scheduler. This allocation can either be continuous in time or in blocks. The utility of the user is defined as a function its throughput. The network utility

---

[2] For simplicity, the BS will also be called the scheduler. In practice, the scheduler could be situated further inside the radio network.

maximization problem is to assign bandwidth or to schedule users in such a way so as to maximize the sum of the utilities of all the users. For proportional fairness, the utility is assumed to be logarithmic.

## 2.1 PF-EXP

Remark that the above problem is a discrete problem, which is not convex yet. Even though the number of options is finite, it is NP-hard to find the optimal solution (see, e.g., [8]). Nevertheless, a simple heuristic, called PF-EXP [7], can be obtained by assuming that the number of users is fixed and that the data rates $r_{i,j}$ are time stationary and ergodic, that is, there is no correlation between $r_{i,j}$ and $r_{i,j+1}$.

The PF-EXP algorithm chooses the user with the highest ratio of the current rate to the observed throughput, that is it chooses the user who is

$$\arg \max_i \frac{r_{i,j}}{\theta_{i,j}}, \tag{1}$$

where $\theta_{i,j} = \beta_j \theta_{i,j-1} + (1 - \beta_j)\alpha_{i,j}r_{i,j}$ the weighted sum allocation of user $i$ until slot $j$. In the long-run when $T$ goes to $\infty$, this algorithm was shown to be optimal for stationary and ergodic channel and fixed number of users [7].

The stationarity assumption is not necessarily true for road traffic when all users always move along a given path. As can be seen in Fig. 1, when users move along the line from left to right, their rate first increases and then decreases. Knowing the position of a vehicle, we can predict the future rate which is not possible when the rate process is stationary and ergodic. Thus, the PF-EXP algorithm need not be optimal for a rate process observed by vehicles.

## 2.2 Predictive Finite-Horizon PF Scheduling ((PF)²S)

In [8], a modified PF algorithm based on predicted future rate was proposed. This algorithms works as follows:

- Predict future rate $\hat{r}_{i,j}$ of cars in every future slot.
- Estimate future allocation $\hat{\alpha}$. They propose three ways to chose and update this $\hat{\alpha}$: (i) *round-robin*, (ii) *blind search* and (iii) *local search*. In our simulations, we use only round-robin because blind search is similar to PF-EXP. Finally, in each time slot, local search iteratively computes until $T$ and then allocates according to (2), making it a computationally expensive method. It is shown in [8] for local search to be effective, the prediction error has to be low for the whole horizon. If the prediction error is high then local search can be worse than round robin.
- For each time slot $j$, compute:

$$M_{i,j} = \frac{r_{i,j}}{\sum_{t=1}^{j-1} \alpha_{i,t}r_{i,t} + \hat{\alpha}_{i,j}r_{i,j} + \sum_{t=j+1}^{T} \hat{\alpha}_{i,t}\hat{r}_{i,t}}. \tag{2}$$

- Choose $i^* = \arg \max_{i \in \{1,2,...K\}} M_{i,j}$.

The index $M_{i,j}$ looks similar to that of the PF-EXP algorithm but includes the future allocation. It is related to the gradient of the utility function in $(I)$. It can be shown that if we can predict correctly $\hat{a}$, then the optimal solution can be obtained.

**Proposition 1.** *If there exist $\alpha^*$ satisfying: $\alpha^*_{i^*,j} = 1$ and $\alpha^*_{i,j} = 0, \forall i \neq i^*_j$, where $i^*_j$ is*

$$\arg \max_{i \in \{1,2,...K\}} \frac{r_{i,j}}{\sum_{t=1}^{j-1} \alpha^*_{i,t} r_{i,t} + \alpha^*_{i,j} r_{i,j} + \sum_{t=j+1} \alpha^*_{i,t} r_{i,t}}. \tag{3}$$

*then $\alpha^*$ is the optimal solution of $(I)$.*

The condition (3) need not always be satisfied. However, when it is, it is sufficient for $\alpha^*$ to be optimal solution of problem $(I)$ as stated in Proposition 1.

For this approach of using the future rates to be efficient, one needs a good estimate of the optimal future throughput. This is not easy because the future throughput is computed from the optimal solution which itself is hard to compute.

## 3   Projected Gradient Approach

We now present a method for estimating the future throughput which improves the utility compared to the $(PF)^2S$ algorithm. The main difficulty of solving $(I)$ is that the problem is discrete. To simplify it, we relax the integer constraints to get a convex optimization problem, which is called relaxed problem as described in $(II)$. The relaxed problem can be solved efficiently using projected gradient based on the projection on simplex formula given in [5].

Consider the following relaxed problem:

$$(II) \begin{cases} \max O(\alpha) = \sum_{i=1}^{K} \log \left( \sum_{j=1}^{T} \alpha_{ij} r_{ij} \right) \\ \text{subject to } \sum_{i=1}^{K} \alpha_{ij} = 1, j = 1, \ldots, T \\ \text{and } \alpha_{ij} \in [0,1] \end{cases}$$

which is very similar to the original problem except that $\alpha_{ij}$ can be non-integer in $[0,1]$. Below, we describe the projected gradient formula for the above relaxed problem.

Denote by $D = \{\alpha \in [0,1]^{K \times T}, \sum_{i=1}^{K} \alpha_{ij} = 1 \, \forall j = 1,2,...,T\}$ the feasible set of the relaxed problem. D is not a simplex yet, therefore we cannot apply directly the algorithm in [5] and we need to modify it.

Denote $\Pi_D \cdot$is the projection on $D$ (see Appendix A). The projected gradient algorithm follows the below steps:

– Initialize $\alpha^0 \in D$ arbitrary.
– From $n = 1, 2, 3, \ldots$ compute: $\alpha^{n+1} = \Pi_D(\alpha^n + \epsilon_n \nabla O(\alpha^n))$, where $\epsilon_n \in (0,1)$ is step size at step $n$.

- Until $\alpha^n$ converges. In our numerical examples, we limited the number of iterations to 20.

Denote by $\tilde{\nabla}O(\alpha) = \Pi_D(\alpha + \epsilon\nabla O(\alpha)) - \alpha$ with the step size $\epsilon \in (0,1)$ small enough. If there is a positive step size $\epsilon$ such that the following condition happens, we have an optimal guarantee of this algorithm, as described the following proposition.

**Proposition 2.** *If $\alpha^* \in D$ and $\tilde{\nabla}O(\alpha^*) = 0$ then $\alpha^*$ is the optimal value of the relaxed problem.*

## 3.1 Projected Gradient Short Term Objective Algorithm (STO1)

Based on the above relaxation, we propose a heuristic which computes the optimal solution for the relaxed problem but at a shorter horizon (Step 1 in the algorithm described below). This is done in order to reduce the computation time of the solution. Further, instead of computing the allocation for each future time slot, we compute the average rate allocated over a larger time slot which corresponds to the time scale at which cars enter and leave the coverage area of the base station (in Step 1). Note that in $\delta = 2$ ms, a car hardly moves any perceivable distance. So, we expect the average channel conditions to change over a much larger time scale (around 1 s) instead of every 2 ms. This larger time scale is also the one in which cars leave and enter the coverage range of the BS. That, is number of cars in the coverage range changes state at this time scale rather than every 2 ms.

At each small time slot $t$, let $a_i(t) = \sum_{j=1}^{t} \alpha_{ij}r_{ij}$ be the cumulative rate of user $i$ until time slot $t$, and $K(t)$ be the number of users inside the coverage range.

Our heuristic algorithm follows the steps:
**Step 1:** In each small slot $t$, we reduce the dimension of variable $\alpha$ and solve the following problem using projected gradient:

$$(III) \begin{cases} \max \sum_{i=1}^{K(t)} U_i \\ \text{subject to } \sum_{i=1}^{K(t)} \alpha_{it} = 1, \\ \text{and } \sum_{i=1}^{K(t)} \bar{\alpha}_{i\tau} = 1, \\ \alpha_{it}, \bar{\alpha}_{i\tau} \in [0,1] \end{cases}$$

where

$$U_i = \log\left(a_i(t-1) + \alpha_{it}r_{it} + \sum_{\tau=1}^{J} \bar{\alpha}_{i\tau}\bar{\rho}_{i\tau}\right),$$

$\tau$ is big slot, $m = \Delta/\delta$, and $\bar{\alpha}_{i\tau}$ is the future allocation in big slot $\tau$. Note that $\alpha_{it}$ and $\bar{\alpha}_{i\tau}$ are the decision variables in problem $(III)$. Also, $\bar{r}_{ij}$ is the average rate in slot $j$ for user $i$, and $\bar{\rho}_{i\tau} = \sum_{j=(\tau-1)m+t+1}^{\tau m+t} \bar{r}_{ij}$, is the total average data rate that user $i$ will experience in big slot $\tau$. The value of $\bar{r}_{ij}$ can be predicted using measurements. We remark that the advantage of using the average value $\bar{r}_{ij}$ instead of the exact value $\hat{r}_{ij}$ as is done in (2) is that the prediction error of an average value will be smaller than that of the exact value.

Since the noise is unpredictable in the future, we assume that only the current rate, $r_{it}$, and the average rate in the future are known.

**Step 2:** In each small slot t, give full allocation for the user who has the *largest allocation computed by* $(\alpha_{it})_{i=\overline{1,K(t)}}$. We observe that: when number of slots is large enough, the optimal is 0-1 almost everywhere, we can show that when time is continuous the solution is 0-1 everywhere (proof omitted due to lack of space). So even in this step round over $(\alpha_{it})_i$, we can hope that we do not go far from the optimal solution of (III).

The complexity of numerically optimal $\alpha$ computation in step 2 is equal to $20(J + 1)\bar{K} \log(\bar{K})$ where 20 is the number of iteration steps of projected gradient in Step 1, $\bar{K}$ is average number of users inside the coverage range, $J$ is the number of big slots.

The proposed algorithm is the similar in spirit to Stochastic Model Predictive Control [9].

## 3.2   Projected Gradient Short Term Objective Algorithm 2 (STO2)

The STO1 algorithm recomputes the future allocation in every small time slot. In order to reduce the computational complexity, in STO2, we propose to recompute the future allocation in every big time slot instead.

**Step 1:** In each big slot $\tau$, we reduce the dimension of variable $\alpha$ and solve the following problem in each big slot by using projected gradient:

$$(III) \begin{cases} \max \sum_{i=1}^{K(\tau)} U_i \\ \text{and } \sum_{i=1}^{K(\tau)} \bar{\alpha}_{i\tau} = 1, \\ \bar{\alpha}_{i\tau} \in [0, 1] \end{cases}$$

where

$$U_i = \log \left( a_i((\tau - 1)m) + \sum_{\tau=1}^{J} \bar{\alpha}_{i\tau} \bar{\rho}_{i\tau} \right).$$

Here $a_i((\tau - 1)m)$ is the total received rate by user $i$ just before the start of big slot $\tau$. The other quantities are same as for algorithm STO1.

**Step 2:** In each small slot $j$ we shall compute $M_{ij}$ as in (2) where the future allocation $\hat{a}$ is the solution $\bar{a}$ of problem $(III)$.

By doing this, we reduce the computation almost $\Delta/\delta$ times since we calculate $\bar{\alpha}$ in each big slot only.

## 4    Numerical Results

We now compare the utility of the proposed heuristic with the PF-EXP, $(PF)^2S$ and a greedy algorithm which allocates to user who has the highest current rate, that is in each time slot $j$ we choose:

$$i^* = \arg \max_{i \in \{1,2,\ldots K\}} r_{ij}.$$

For the $(PF)^2S$ the future allocation was done using the round robin algorithm.
    Denote by

$$O^A = \sum_{i=1}^{K} \log \Big( \sum_{j=1}^{T} \alpha_{ij}^A r_{ij} \Big),$$

the total reward of algorithm $A$ and by $\bar{O}^A = \frac{1}{K} O^A$ its average reward over $K$ users. Given $A, B$ two algorithms, then the ratio between $A, B$ equals $\exp(\bar{O}^A - \bar{O}^B)$. The percentage of improvement of algorithm $A$ over $B$ is computed equal to $(\exp(\bar{O}^A - \bar{O}^B) - 1) \cdot 100\%$.
    Due to the logarithm in the objective function, taking a different unit of measure for the rate will give a different percentage improvement between algorithms. Although logarithm is an increasing function, we can know which algorithm is better than the other, but we will not get a consistent percentage improvement across different units of measure. Therefore, by taking the difference as above we construct a consistent criterion for comparison.
    The road length is taken to be $L = 1000$ m with 0 at the leftmost edge. The closest point on the road to the BS is at $x = 500$ m. The data rate at position $x$ along the road is given by:

$$r(x) = \eta \cdot (1 + \kappa \exp (|x - 500|/\sigma) , \tag{4}$$

where $\kappa \geq 0$ is a real number and $\eta$ is uniform random variable whose range will be in $[0.8, 1.2]$ unless stated otherwise. A sample path of $r(x)$ is shown in Fig. 2. This function has the highest mean at the mid-point of the segment and the lowest mean at the two end points.
    We emphasize the algorithm itself is independent of the rate function. We chose the above rate function for convenience.
    The time horizon $T$ was 4000000 small time slots which is 8000 secs or a little over two hours. The big slot length $\Delta$ for our projected gradient short term objective algorithm was taken as 1 s or equivalently 500 small time slots.

### 4.1    Homogeneous Vehicle Velocities

First, we show the results when all vehicles move with the same velocity which is taken to be $v = 25$ m/s. That is, there are $N = 20000$ spatial small slots in the coverage range and $J = 40$ big slots. A new car enters through the left edge in every big slot (i.e., every second) with probability $p$.

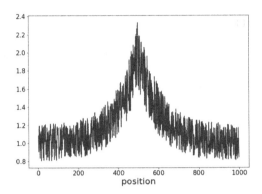

**Fig. 2.** Sample path of data rate at various positions along the road. $\sigma = 100, \kappa = 1$ and $\eta \in [0.8, 1.2]$.

Figure 3 shows the average utility obtained by a vehicle for each of the four algorithms as a function of the probability of arrival of car in each big slot. Figure 4 shows the percentage improvement of three other algorithms compared to PF-EXP. The proposed algorithm does better than PF-EXP and more importantly better than $(PF)^2S$. Although, we have shown the greedy algorithm for comparison, we emphasize that greedy is not practically implemented because it can be very unfair to users that have heterogeneous rates. In the simulated scenario, all vehicles move along the same road and observe statistically identical but position dependent conditions during their stay. These conditions are rather favorable for the greedy algorithm.

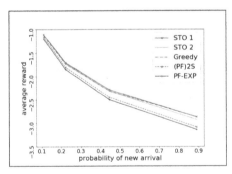

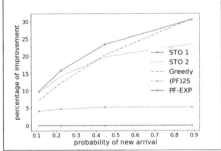

**Fig. 3.** Average reward per car. Homogeneous velocities.

**Fig. 4.** Percentage improvement over PF-EXP. Homogeneous velocities.

## 4.2 Comparison with the Upper Bound

Next, again for homogeneous velocities, we also include the solution of the relaxed problem $(II)$ but for a smaller road length and shorter horizon because it is computationally expensive. The parameters for this setting are: $L = 100\,\text{m}$, $J = 40$ big slots, $T = 500\,\text{s}$, and the other parameters are the same as in the homogeneous case. We assume that the relaxed algorithm knows the future arrivals and the future rate exactly whereas the other algorithms do not know this information. The solution to the relaxed problem gives an upper bound to the optimal solution of the original problem in $(I)$.

Figures 5 and 6 plot the average reward per car and percentage improvement for the five algorithms with respect to PF-EXP. It is seen that the proposed algorithm is quite close to the upper bound in this scenario.

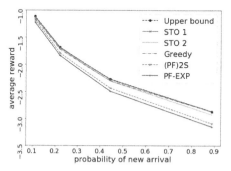 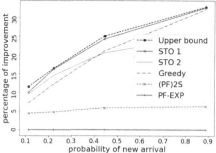

**Fig. 5.** Average reward per car. Includes the upper bound from the solution of $(II)$. Small setting of homogeneous velocities.

**Fig. 6.** Percentage improvement over PF-EXP. Small setting of homogeneous velocities.

## 4.3 Heterogeneous Vehicle Velocities

Finally, we show the results when a fraction $q$ of vehicles move with $v_0 = 25\,\text{m/s}$ and the other fraction with $v_1 = 12.5\,\text{m/s}$. Note that the proposed algorithm takes the larger of the two values of $N$ computed with the two velocities. The faster class of cars will be called class 0. Here, the horizon, $T$, is a little over 2 h.

For this scenario, the probability of new a arrival is fixed at $p = 2/9$ in Fig. 7, 8 and at $p = 4/9$ in Fig. 9, 10. Thus a new car of class 0 arrives with probability $p \cdot q$ and with probability $p(1 - q)$ a new car of class 1 arrives. A new car enters through the left edge in every big slot (i.e., every second) with probability $p$.

Figure 7 (resp. Fig. 9) shows the average utility obtained by a vehicle for each of the four algorithms as a function $q$ for probability of new arrival $p = 2/9$ (resp. $p = 4/9$). Figure 8 (resp. Fig. 10) shows the percentage improvement of three other algorithms compared to PF-EXP for the two probabilities of arrival as before. As before, the proposed algorithm does better than both PF and $(PF)^2S$.

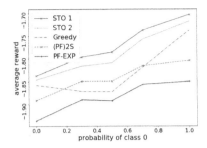

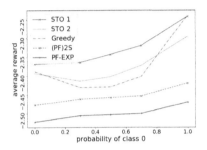

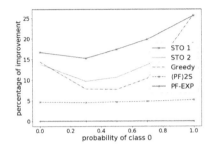

**Fig. 7.** Average reward per car. Heterogeneous velocities. $p = 2/9$.

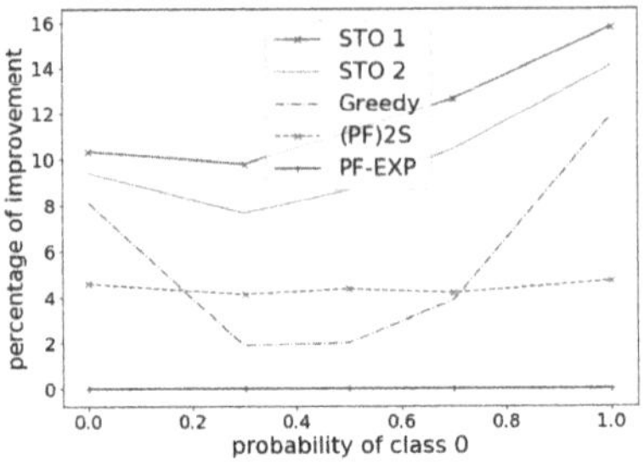

**Fig. 8.** Percentage improvement over PF-EXP. Heterogeneous velocities. $p = 2/9$.

**Fig. 9.** Average reward per car. Heterogeneous velocities. $p = 4/9$.

**Fig. 10.** Percentage improvement. Heterogeneous velocities. $p = 4/9$.

## 5  Future Work

The results in this paper were obtained on a simplified scenario of one base station serving only cars traveling on one road. Our immediate goal is to evaluate the performance of the proposed algorithm in more complex scenarios with a network of multiple base stations and different classes or mobile (e.g., pedestrians) and stationary users. The algorithm will be designed to be run in either a coordinated or a distributed manner.

Other directions of research include: investigation of the tradeoff between the reduction of complexity (e.g., by changing the size of the horizon and the big slots) and the quality of the heuristic; integration of users with different classes of QoS requirements including latency, jitter, or periodic communication constraints; and computation of analytical bounds on the sub optimality of the proposed algorithm.

# A    Projection on Feasible Set $D$

In fact $D$ is a Cartesian product of $J$ simplexes: $D = D_1 \times D_2 \cdots \times D_J$ where

$$D_j = \{a_j = (\alpha_{ij})_{i=\overline{1,K}} \in [0,1]^K, \sum_{i=1}^{K} \alpha_{ij} = 1\}$$

for all $j = 1, 2, ..., J$.
Then $(D_j)_j$ are simplexes, so we can compute the projection on $D_j$ following [5]. The projection on D can be computed by the simple following lemma:

**Lemma 1.** *If* $Y = (y_{ij})_{i=\overline{1,K}, j=\overline{1,J}} \in \mathbb{R}^{K \times J}$, *then*

$$\Pi_D(Y) = \Pi_{D_1}(Y_1) \times \Pi_{D_2}(Y_2) \times \cdots \times \Pi_{D_J}(Y_J),$$

*where* $Y_j = (y_{ij})_{i=\overline{1,K}}.$

*Proof.* (of the Lemma 1) Denote by $Z = \Pi_{D_1}(Y_1) \times \Pi_{D_2}(Y_2) \times \cdots \times \Pi_{D_J}(Y_J)$. It is obvious to check that for any $X \in D^{K \times J}$ then $\langle Y - Z, X - Z \rangle \leq 0$.

As describe in [5], the complexity of finding $\Pi_{D_j}$ is equal to $K \log(K)$ by observation in practice, and equal to $O(K^2)$ in the worst case. Therefore the complexity of finding projection on $D = D_1 \times D_2 \cdots \times D_J$ is equal to $JK \log(K)$ in practice.

# B    Modeling as an Optimal Control Problem

As aforementioned, in this part we prove 0-1 everywhere property of the solution of the continuous time for the original problem.

When user $i$ moves in the road, his position changes continuously, denote by $x_i(t)$ position of that user. The rate changes in the way we explain above according to position of the user, denoted by $r_i(x_i(t))$. Assume that the BS can allocate in continuous time, we denote by $a_i(t)$ is the allocation for user $i$ at time $t$, we relaxed the integer constraint so that $a_i(t) \in [0,1]$ then we get a continuous control optimization:

$$(VI) \begin{cases} \max \sum_{i=1}^{K} \log \left( \frac{1}{T} \int_0^T \alpha_i(t) r(t, x_i(t)) dt \right) \\ \text{such that } \alpha_i \in \mathcal{S}^K \forall t, \forall i, \end{cases}$$

On other hand we have Mayer form with the *terminal cost function* as follows:

$$\begin{cases} \max \sum_{i=1}^{K} \log(y_i(T)), \\ \dot{y}_i(t) = \alpha_i(t) r(t, x_i(t)) dt, y_i(0) = 0 \forall i, \\ \dot{x}(t) = v(t) \forall i, \end{cases}$$

We can solve (VI) to get the optimal solution by using maximum principle. The solution of (VI) can be not unique, but one of its solution has the integer form as described in the following proposition:

**Proposition 3.** *The solution of the problem (I) is of the form: for every* $t$, $\alpha_{i^*}(t) = 1$ *and* $\alpha_j(t) = 0$ *for all* $j \neq i^*(t)$, *where*

$$i^*(t) = \arg\max_i \frac{r(t, x_i(t))}{y_i(T)}.$$

## C    Proofs

*Proof.* (proof of Proposition 2) The optimal is obtained by proving that for any $\alpha \in D$,

$$\nabla O(\alpha^*)(\alpha^* - \alpha) \le 0.$$

Because of Lemma 1, we will reduce the proof on $D_1$. Assuming $O$ is convex function on $D_1$, we shall prove that if $\alpha^* = (\alpha_i^*)_{i=1,\ldots,K} \in D_1$ satisfies

$$\Pi_{D_1}(\alpha^* + \epsilon \nabla(\alpha^*)) = \alpha^* \tag{5}$$

where $\epsilon$ positive, then

$$\nabla O(\alpha^*)(\alpha^* - \alpha) \ge 0, \text{ for any } \alpha \in D_1$$

i.e, $\alpha^*$ is global optimal of $O$. Indeed, without loss of generality, we assume that

$$\alpha_1^* + \epsilon \frac{\partial O}{\partial \alpha_1^*} \ge \alpha_2^* + \epsilon \frac{\partial O}{\partial \alpha_2^*} \ge \ldots \ge +\alpha_M^* + \epsilon \frac{\partial O}{\partial \alpha_M^*} \ge \ldots \ge \alpha_K^* + \epsilon \frac{\partial O}{\partial \alpha_K^*}$$

where $M$ is the largest index such that

$$\frac{1}{M} \sum_{i=1}^{M} (\alpha_i^* + \epsilon \frac{\partial O}{\partial \alpha_i^*} - 1) \le \alpha_M^* + \epsilon \frac{\partial O}{\partial \alpha_M^*}.$$

Denote by $\tau = \frac{1}{M} \sum_{i=1}^{M} (\alpha_i^* + \epsilon \frac{\partial O}{\partial \alpha_i^*} - 1)$, by proposition 10 in [5] we have:
$$\Pi_{D_1}(\alpha^* + \epsilon \cdot \nabla(\alpha^*)) = (\alpha_1^* + \epsilon \frac{\partial O}{\partial \alpha_1^*} - \tau, \alpha_2^* + \epsilon \frac{\partial O}{\partial \alpha_2^*} - \tau, \ldots, \alpha_M^* + \epsilon \frac{\partial O}{\partial \alpha_M^*} - \tau, 0, \ldots, 0).$$
Using (5) to compare term by term we get:

1. $\alpha_{M+1}^* = \cdots = \alpha_K^* = 0$,
2. $\alpha_{M+1}^* + \epsilon \frac{\partial O}{\partial \alpha_{M+1}^*} \le \tau, \cdots, \alpha_K^* + \epsilon \frac{\partial O}{\partial \alpha_K^*} \le \tau$. Now, from the first item we have
   $\alpha_{M+1}^* = \cdots = \alpha_K^* = 0$. It implies $\epsilon \frac{\partial O}{\partial \alpha_{M+1}^*} \le \tau, \cdots, \epsilon \frac{\partial O}{\partial \alpha_K^*} \le \tau$,
3. $\epsilon \frac{\partial O}{\partial \alpha_1^*} = \cdots = \epsilon \frac{\partial O}{\partial \alpha_M^*} = \tau.$

Thus,

$$
\epsilon \nabla O(\alpha^*)(\alpha^* - \alpha) = \sum_{i=1}^{K} \epsilon \frac{\partial O}{\partial \alpha_i^*}(\alpha_i^* - \alpha_i)
$$

$$
= \sum_{i=1}^{M} \epsilon \frac{\partial O}{\partial \alpha_i^*}(\alpha_i^* - \alpha_i) + \sum_{i=M+1}^{K} \epsilon \frac{\partial O}{\partial \alpha_i^*}(\alpha_i^* - \alpha_i),
$$

$$
= \sum_{i=1}^{M} \tau(\alpha_i^* - \alpha_i) + \sum_{i=M+1}^{K} \epsilon \frac{\partial O}{\partial \alpha_i^*}(\alpha_i^* - \alpha_i),
$$

$$
= \sum_{i=1}^{K} \tau \alpha_i^* - \sum_{i=1}^{K} \tau \alpha_i + \sum_{i=M+1}^{K} \left(\epsilon \frac{\partial O}{\partial \alpha_i^*} - \tau\right)(\alpha_i^* - \alpha_i),
$$

$$
= \tau - \tau + \sum_{i=M+1}^{K} \left(\epsilon \frac{\partial O}{\partial \alpha_i^*} - \tau\right)(0 - \alpha_i)
$$

$$
\geq 0.
$$

The last sum less than 0 since all its terms are greater than or equal to 0.

*Proof.* (proof of Proposition 1) In fact the condition (3) implies that $\tilde{\nabla} O(\alpha^*) = 0$ and from Proposition 2 we have conclusion.

# References

1. Bang, H.J., Ekman, T., Gesbert, D.: Channel predictive proportional fair scheduling. IEEE Trans. Wireless Commun. **7**(2), 482–487 (2008)
2. Borst, S.: User-level performance of channel-aware scheduling algorithms in wireless data networks. IEEE/ACM Trans. Netw. **13**(3), 636–647 (2005)
3. Borst, S., Hegde, N., Proutiere, A.: Mobility-driven scheduling in wireless networks. In: IEEE INFOCOM 2009, pp. 1260–1268 (2009)
4. Chandur, P., Karthik, R.M., Sivalingam, K.M.: Performance evaluation of scheduling algorithms for mobile WiMAX networks. In 2012 IEEE International Conference on Pervasive Computing and Communications Workshops, March 2012, pp. 764–769 (2012)
5. Condat, L.: Fast projection onto the simplex and the l1 ball. Math. Program. **158**, 575–585 (2016)
6. Kelly, F.: Charging and rate control for elastic traffic. Eur. Trans. Telecommun. **8**(1), 33–37 (1997)
7. Kushner, H.J., Whiting, P.A.: Convergence of proportional-fair sharing algorithms under general conditions. IEEE Trans. Wireless Commun. **3**(4), 1250–1259 (2004)
8. Margolies, R., et al.: Exploiting mobility in proportional fair cellular scheduling: measurements and algorithms. IEEE/ACM Trans. Netw. **24**(1), 355–367 (2016)
9. Mesbah, A.: Stochastic model predictive control: an overview and perspectives for future research. IEEE Control Syst. Mag. **36**(6), 30–44 (2016)
10. Mo, J., Walrand, J.: Fair end-to-end window-based congestion control. IEEE/ACM Trans. Netw. **8**(5), 556–567 (2000)

11. Tan, L., Zhu, Z., Ge, F., Xiong, N.: Utility maximization resource allocation in wireless networks: methods and algorithms. IEEE Trans. Syst. Man Cybern.: Syst. **45**(7), 1018–1034 (2015)
12. Yi, Y., Chiang, M.: Stochastic network utility maximisation-a tribute to Kelly's paper published in this journal a decade ago. Eur. Trans. Telecommun. **19**(4), 421–442 (2008)
13. Zhou, H., Fan, P., Li, J.: Global proportional fair scheduling for networks with multiple base stations. IEEE Trans. Veh. Technol. **60**(4), 1867–1879 (2011)

# Method of Asymptotic Diffusion Analysis of Queueing System $M|M|N$ with Feedback

Anatoly Nazarov$^{(\boxtimes)}$ ⓘ, Svetlana Paul$^{(\boxtimes)}$ ⓘ, and Ekaterina Pavlova$^{(\boxtimes)}$ ⓘ

Tomsk State University, Lenin av. 36, Tomsk, Russia
nazarov.tsu@gmail.com, paulsv82@mail.ru, pavlovakatya_2010@mail.ru

**Abstract.** In this article we consider a mathematical model of re-servicing customers in the form of a queuing system consisting of N servers with feedback and orbit. Incoming flow is the Poisson flow of customers. To obtain probability distribution of the number of customers in the orbit we use method of asymptotic diffusion analysis.

**Keywords:** Queueing system · Server · Feedback · Orbit · Asymptotic diffusion analysis

## 1 Introduction

In practice queuing systems in which served calls require re-service [1–6] (depending on the quality of service received, external factors, and so on) are quite common. Such situations are often encountered in Multi-Agent Systems (MAS), where requests that receive satisfactory service are re-addressed to the same agents. The functioning of multi-agent systems is accurately described by models of queuing systems with feedback. The application of queuing theory to optimize the performance of MAS is described in detail in [7,8].

The difference from the already known studies of queuing systems with repeated calls is that the orbit is formed from calls that have already received service [9,10].

It is known that for the described systems it is not possible to obtain an explicit form of distribution or it is difficult to develop methods to obtain it in systems with more than 2 channels. That is why the development of new approaches is so relevant. Besides note that retrial queuing systems with an orbit consisting of serviced requests have not been sufficiently studied.

In this paper we propose to find an approximate form of formulas by the new original method of asymptotic diffusion analysis, which is development of the classical methods of asymptotic analysis [11–14].

This study (research grant No 8.1.16.2019) was supported by The Tomsk State University competitiveness improvement programme.

M. Gribaudo et al. (Eds.): ASMTA 2019, LNCS 12023, pp. 131–143, 2020.
https://doi.org/10.1007/978-3-030-62885-7_10

## 2    Statement of the Problem and Mathematical Model

Consider the queueing system with N servers with feedback (Fig. 1). Customers arrive in the system according to a Poisson process with parameter $\lambda$.

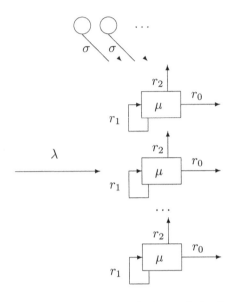

**Fig. 1.** Queueing system $M|M|N$ with feedback

At the time of occurrence customer in the system it goes to a free server where its service is performed during a random time having exponential distribution function with parameter $\mu$. If there are no free servers in the system at the moment of occurrence, then customer goes to the orbit instantly and it carries out a delay there for a random time having exponential distribution function with parameter $\sigma$ [15–18].

At the moment of completion of service customer leaves the system with probability $r_0$; goes to re-service implementing feedback instantly with probability $r_1$; goes to the orbit implementing delayed feedback with probability $r_2$, where it stayed for a random time having exponential distribution function with parameter $\sigma$ and after then requires service again.

Let us denote by $n(t)$ – the number of occupied servers in the system at the moment $t, n = 0, N, i(t)$ – the number of customers in the orbit at the moment $t$.

Set the problem of obtaining probability distribution of the number of occupied servers and the number of customers in the orbit.

## 3    Kolmogorov Differential Equations

Let us consider two-dimensional Markov process $\{n(t), i(t)\}$ and for its probability distribution $P\{n(t) = n, i(t) = i\} = P(n, i, t)$ we can write the system of

Kolmogorov differential equations

$$\frac{\partial P(n,i,t)}{\partial t} = -(\lambda + n\mu + i\sigma)P(n,i,t)$$
$$+ \lambda P(n-1,i,t) + (i+1)\sigma P(n-1,i+1,t) + (n+1)r_0\mu P(n+1,i,t)$$
$$+ (n+1)r_2\mu P(n+1,i-1,t) + nr_1\mu P(n,i,t), \qquad (1)$$
$$\frac{\partial P(N,i,t)}{\partial t} = -(\lambda + N\mu)P(N,i,t) + \lambda P(N-1,i,t)$$
$$+ \lambda P(N,i-1,t) + (i+1)\sigma P(N-1,i+1,t) + Nr_1\mu P(N,i,t).$$

We introduce the partial characteristic function [12]

$$H(n,u,t) = \sum_{i=0}^{\infty} e^{jui} P(n,i,t),$$

where $j = \sqrt{-1}$ is the imaginary unit. Then we can rewrite system (1):

$$\frac{\partial H(n,u,t)}{\partial t} = -(\lambda + n\mu(1-r_1))H(n,u,t)$$
$$+ \lambda H(n-1,u,t) + (n+1)\mu(r_0 + r_2 e^{ju})H(n+1,u,t)$$
$$- j\sigma e^{-ju}\frac{\partial H(n-1,u,t)}{\partial u} + j\sigma\frac{\partial H(n,u,t)}{\partial u}, \qquad (2)$$
$$\frac{\partial H(N,u,t)}{\partial t} = (\lambda(e^{ju}-1) - N\mu(1-r_1))H(N,u,t)$$
$$+ \lambda H(N-1,u,t) + j\sigma e^{-ju}\frac{\partial H(N-1,u,t)}{\partial u}.$$

Denote row-vectors

$$\mathbf{H}(u,t) = \{H(0,u,t), H(1,u,t), ..., H(N,u,t)\},$$

$$\frac{\partial \mathbf{H}(u,t)}{\partial t} = \left\{\frac{\partial H(0,u,t)}{\partial t}, \frac{\partial H(1,u,t)}{\partial t}, ...., \frac{\partial H(N,u,t)}{\partial t}\right\},$$

$$\frac{\partial \mathbf{H}(u,t)}{\partial u} = \left\{\frac{\partial H(0,u,t)}{\partial u}, \frac{\partial H(1,u,t)}{\partial u}, ...., \frac{\partial H(N,u,t)}{\partial u}\right\}$$

we rewrite Eqs. (2) in matrix form

$$\frac{\partial \mathbf{H}(u,t)}{\partial t} = \mathbf{H}(u,t)\left(\mathbf{A} + e^{ju}\mathbf{B}\right) + j\sigma\frac{\partial \mathbf{H}(u,t)}{\partial u}\left(\mathbf{I_0} - e^{-ju}\mathbf{I_1}\right),$$

where

$$\mathbf{A} = \begin{bmatrix} -\lambda & \lambda & 0 & \cdots & 0 \\ \mu r_0 & -(\lambda + \mu(r_0 + r_2)) & \lambda & \cdots & 0 \\ 0 & 2\mu r_0 & -(\lambda + 2\mu(r_0 + r_2)) & \cdots & 0 \\ \cdots & \cdots & \cdots & \cdots & \cdots \\ 0 & 0 & 0 & \cdots & -(\lambda + N\mu(r_0 + r_2)) \end{bmatrix},$$

$$
\mathbf{B} = \begin{bmatrix} 0 & 0 & ... & 0 & 0 \\ \mu r_2 & 0 & ... & 0 & 0 \\ 0 & 2\mu r_2 & ... & 0 & 0 \\ ... & ... & ... & ... & ... \\ 0 & 0 & ... & 0 & 0 \\ 0 & 0 & ... & N\mu r_2 & \lambda \end{bmatrix}, \mathbf{I_0} = \begin{bmatrix} 1 & 0 & ... & 0 & 0 \\ 0 & 1 & ... & 0 & 0 \\ ... & ... & ... & ... & ... \\ 0 & 0 & ... & 1 & 0 \\ 0 & 0 & ... & 0 & 0 \end{bmatrix}, \mathbf{I_1} = \begin{bmatrix} 0 & 1 & 0 & ... & 0 & 0 \\ 0 & 0 & 1 & ... & 0 & 0 \\ ... & ... & ... & ... & ... & ... \\ 0 & 0 & 0 & ... & 0 & 1 \\ 0 & 0 & 0 & ... & 0 & 0 \end{bmatrix}.
$$

We multiply matrix equation by column-vector of units $\mathbf{e}$ taking into account $(\mathbf{A} + \mathbf{B})\mathbf{e} = 0$ and $(\mathbf{I_0} - \mathbf{I_1})\mathbf{e} = 0$, we obtain

$$
\frac{\partial \mathbf{H}(u,t)}{\partial t}\mathbf{e} = \mathbf{H}(u,t)(e^{ju} - 1)\mathbf{B}\mathbf{e} + j\sigma e^{-ju}\frac{\partial \mathbf{H}(u,t)}{\partial u}(e^{ju} - 1)\mathbf{I_0}\mathbf{e}
$$

$$
= (e^{ju} - 1)\Big\{\mathbf{H}(u,t)\mathbf{B}\mathbf{e} + j\sigma e^{-ju}\frac{\partial \mathbf{H}(u,t)}{\partial u}\mathbf{I_0}\mathbf{e}\Big\}.
$$

Thus, matrix system (2) and scalar equation have form

$$
\frac{\partial \mathbf{H}(u,t)}{\partial t} = \mathbf{H}(u,t)\big(\mathbf{A} + e^{ju}\mathbf{B}\big) + j\sigma\frac{\partial \mathbf{H}(u,t)}{\partial u}\big(\mathbf{I_0} - e^{-ju}\mathbf{I_1}\big),
$$

$$
\frac{\partial \mathbf{H}(u,t)}{\partial t}\mathbf{e} = (e^{ju} - 1)\Big\{\mathbf{H}(u,t)\mathbf{B}\mathbf{e} + j\sigma e^{-ju}\frac{\partial \mathbf{H}(u,t)}{\partial u}\mathbf{I_0}\mathbf{e}\Big\}. \tag{3}
$$

We will get solution of the problem (3) by asymptotic diffusion analysis [19] under condition of a large delay of customers in the orbit ($\sigma \to 0$) [20].

## 4    Asymptotic Diffusion Analysis Method

Denote by $\sigma = \varepsilon$ and make the following substitutions

$$
\tau = t\varepsilon, u = \varepsilon w, \mathbf{H}(u,t) = \mathbf{F}(w,\tau,\varepsilon). \tag{4}
$$

Then we can rewrite (3)

$$
\varepsilon\frac{\partial \mathbf{F}(w,\tau,\varepsilon)}{\partial \tau} = \mathbf{F}(w,\tau,\varepsilon)\big(\mathbf{A} + e^{j\varepsilon w}\mathbf{B}\big) + j\frac{\partial \mathbf{F}(w,\tau,\varepsilon)}{\partial w}\big(\mathbf{I_0} - e^{-j\varepsilon w}\mathbf{I_1}\big),
$$

$$
\varepsilon\frac{\partial \mathbf{F}(w,\tau,\varepsilon)}{\partial \tau}\mathbf{e} = (e^{j\varepsilon w} - 1)\Big\{\mathbf{F}(w,\tau,\varepsilon)\mathbf{B}\mathbf{e} + je^{-j\varepsilon w}\frac{\partial \mathbf{H}(u,t)}{\partial u}\mathbf{I_0}\mathbf{e}\Big\}. \tag{5}
$$

Solving system (5) under condition $\varepsilon \to 0$, we prove the following statement.

**Theorem 1.** *In the considered feedback system the components $R(n)$ of the probability distribution vector of the number of occupied servers $\mathbf{R}$ are determined by the equations*

$$
R(n) = \frac{\lambda + x}{n\mu(r_0 + r_2)}R(n - 1) = \Big(\frac{\lambda + x}{\mu(r_0 + r_2)}\Big)^n\frac{1}{n!}R(0),
$$

$$
R(0) = 1/\sum_{i=0}^{N}\Big(\frac{\lambda + x}{\mu(r_0 + r_2)}\Big)^n\frac{1}{n!}, \tag{6}
$$

where row-vector $\mathbf{R} = \{R(0), R(1), ..., R(N)\}$ – the probability distribution of the number of occupied servers in the system and $x$ is a solution to the equation $x = x(\tau)$: $x'(\tau) = \mathbf{R}\mathbf{B}\mathbf{e} - x(\tau)\mathbf{R}\mathbf{I_0}\mathbf{e}$, which is determined below.

Proof. We consider the first equation of system (5) in limit $\varepsilon \to 0$, denote $\lim_{\varepsilon \to 0} \mathbf{F}(w, \tau, \varepsilon) = \mathbf{F}(w, \tau)$ and obtain

$$\mathbf{F}(w, \tau)(\mathbf{A} + \mathbf{B}) + j\frac{\partial \mathbf{F}(w, \tau)}{\partial w}(\mathbf{I_0} - \mathbf{I_1}) = 0. \tag{7}$$

We find solution of the problem (7) in the form $\mathbf{F}(w, \tau) = \mathbf{R}e^{jwx(\tau)}$, we define the function $x(\tau)$ below, then we obtain the following system

$$\mathbf{R}\Big\{(\mathbf{A} + \mathbf{B}) - x(\tau)(\mathbf{I_0} - \mathbf{I_1})\Big\} = 0$$
$$\mathbf{Re} = 1. \tag{8}$$

Taking into account $(\mathbf{A} + \mathbf{B})\mathbf{e} = 0$ and $(\mathbf{I_0} - \mathbf{I_1})\mathbf{e} = 0$, we know that the determinant of the matrix $\mathbf{A} + \mathbf{B} - x(\tau)(\mathbf{I_0} - \mathbf{I_1})$ is 0. Write first equation of system (8) in scalar form

$$(\lambda + x)R(n-1) - \{\lambda + x + n\mu(r_0 + r_2)\}R(n) + (n+1)\mu(r_0 + r_2)R(n+1) = 0.$$

Let us denote $(\lambda + x)R(n-1) - n\mu(r_0 + r_2)R(n) = z_{n-1}$, then $z_{n-1} - z_n = 0$. We obtain

$$R(n) = \frac{\lambda + x}{n\mu(r_0 + r_2)}R(n-1) = \left(\frac{\lambda + x}{\mu(r_0 + r_2)}\right)^n \frac{1}{n!}R(0),$$
$$R(0) = 1/\sum_{i=0}^{N}\left(\frac{\lambda + x}{\mu(r_0 + r_2)}\right)^n \frac{1}{n!}.$$

Remark to the Theorem. From the Eq. (6) follows that for any positive value of $x$, the probability distribution $R(n)$ is a discrete Erlang distribution with the parameter $\gamma = \frac{\lambda + x}{\mu(r_0 + r_2)}$.

Let us find $x = x(\tau)$. Consider second equation of the system (5) in limit $\varepsilon \to 0$

$$\frac{\partial \mathbf{F}(w, \tau)}{\partial \tau}\mathbf{e} = jw\Big\{\mathbf{F}(w, \tau)\mathbf{B}\mathbf{e} + j\frac{\partial \mathbf{F}(w, \tau)}{\partial w}\mathbf{I_0}\mathbf{e}\Big\},$$

and substitute solution $\mathbf{F}(w, \tau) = \mathbf{R}e^{jwx(\tau)}$, then

$$x'(\tau) = \mathbf{R}\mathbf{B}\mathbf{e} - x(\tau)\mathbf{R}\mathbf{I_0}\mathbf{e}. \tag{9}$$

The stationary solution of Eq. (9) is a positive solution $x = \kappa$ to the scalar equation $\mathbf{R}(\mathbf{B} - x\mathbf{I_0})\mathbf{e} = 0$.

Substituting $x = \kappa$ to (6), we obtain stationary probability distribution $\mathbf{R}$ of number of occupied servers. Here $\kappa$ – positive root of equation $x'(\tau) = 0$.

Denote by

$$x'(\tau) = a(x) = \mathbf{R}\mathbf{B}\mathbf{e} - x(\tau)\mathbf{R}\mathbf{I_0}\mathbf{e}. \tag{10}$$

We make the following substitutions in (3)

$$\mathbf{H}(u,t) = e^{j\frac{u}{\sigma}x(\sigma t)}\mathbf{H}^{(1)}(u,t),$$

obtain the system

$$\frac{\partial \mathbf{H}^{(1)}(u,t)}{\partial t} + jux'(\sigma t)\mathbf{H}^{(1)}(u,t) = \mathbf{H}^{(1)}(u,t)(\mathbf{A} + e^{ju}\mathbf{B})$$
$$+ j\sigma\left[\frac{j}{\sigma}x(\sigma t)\mathbf{H}^{(1)}(u,t) + \frac{\partial \mathbf{H}^{(1)}(u,t)}{\partial u}\right](\mathbf{I_0} - e^{-ju}\mathbf{I_1}),$$
$$\left[\frac{\partial \mathbf{H}^{(1)}(u,t)}{\partial t} + jux'(\sigma t)\mathbf{H}^{(1)}(u,t)\right]\mathbf{e}$$
$$= (e^{ju} - 1)\left\{\mathbf{H}^{(1)}(u,t)\mathbf{Be} + j\sigma e^{-ju}\left[\frac{j}{\sigma}x(\sigma t)\mathbf{H}^{(1)}(u,t) + \frac{\partial \mathbf{H}^{(1)}(u,t)}{\partial u}\right]\mathbf{I_0e}\right\}. \tag{11}$$

Taking into account (10) we rewrite system (11)

$$\frac{\partial \mathbf{H}^{(1)}(u,t)}{\partial t} + jua(x)\mathbf{H}^{(1)}(u,t) = \mathbf{H}^{(1)}(u,t)(\mathbf{A} + e^{ju}\mathbf{B} - x(\mathbf{I_0} - e^{-ju}\mathbf{I_1}))$$
$$+ j\sigma\frac{\partial \mathbf{H}^{(1)}(u,t)}{\partial u}(\mathbf{I_0} - e^{-ju}\mathbf{I_1}),$$
$$\frac{\partial \mathbf{H}^{(1)}(u,t)}{\partial t}\mathbf{e} + jua(x)\mathbf{H}^{(1)}(u,t)\mathbf{e}$$
$$= (e^{ju} - 1)\left(\mathbf{H}^{(1)}(u,t)[\mathbf{Be} - e^{-ju}x\mathbf{I_0}] + e^{-ju}j\sigma\frac{\partial \mathbf{H}^{(1)}(u,t)}{\partial u}\mathbf{I_0}\right)\mathbf{e}. \tag{12}$$

Denote by $\sigma = \varepsilon^2$ and make following substitutions in (12)

$$\tau = t\varepsilon^2, u = \varepsilon w, \mathbf{H}^{(1)}(u,t) = \mathbf{F}^{(1)}(w,\tau,\varepsilon) \tag{13}$$

we can write

$$\varepsilon^2\frac{\partial \mathbf{F}^{(1)}(w,\tau,\varepsilon)}{\partial \tau} + j\varepsilon wa\mathbf{F}^{(1)}(w,\tau,\varepsilon) = \mathbf{F}^{(1)}(w,\tau,\varepsilon)(\mathbf{A} + e^{j\varepsilon w}\mathbf{B} - x(\mathbf{I_0} - e^{-j\varepsilon w}\mathbf{I_1}))$$
$$+ j\varepsilon\frac{\partial \mathbf{F}^{(1)}(w,\tau,\varepsilon)}{\partial w}(\mathbf{I_0} - e^{-j\varepsilon w}\mathbf{I_1}),$$
$$\varepsilon^2\frac{\partial \mathbf{F}^{(1)}(w,\tau,\varepsilon)}{\partial \tau}\mathbf{e} + j\varepsilon wa\mathbf{F}^{(1)}(w,\tau,\varepsilon)\mathbf{e}$$
$$= (e^{j\varepsilon w} - 1)\left(\mathbf{F}^{(1)}(w,\tau,\varepsilon)[\mathbf{Be} - e^{-j\varepsilon w}x\mathbf{I_0}] + e^{-j\varepsilon w}j\varepsilon\frac{\partial \mathbf{F}^{(1)}(w,\tau,\varepsilon)}{\partial w}\mathbf{I_0}\right)\mathbf{e}. \tag{14}$$

We prove the following statement.

**Theorem 2.** *The probability distribution density of the normalized number of calls in orbit has the form*

$$\pi(z) = \frac{C}{b(z)}exp\left\{\frac{2}{\sigma}\int_0^z \frac{a(x)}{b(x)}dx\right\}, \tag{15}$$

*where* $C$ – *constant,*

$$a(x) = \mathbf{RBe} - x\mathbf{RI_0e},$$
$$b(x) = 7a(x) + 2\mathbf{g}[\mathbf{B} - x\mathbf{I_0}]\mathbf{e} + 2\mathbf{R}x\mathbf{I_0e}, \tag{16}$$

*here row-vector* $\mathbf{g}$ *is determined by the system of equations*

$$\mathbf{g}(\mathbf{A} + \mathbf{B} + x(\mathbf{I_1} - \mathbf{I_0})) = a\mathbf{R} + \mathbf{R}(x\mathbf{I_1} - \mathbf{B}),$$
$$\mathbf{ge} = 0. \tag{17}$$

*Proof.* We write the first Eq. (14) up to $O(\varepsilon^2)$:

$$j\varepsilon wa\mathbf{F}^{(1)}(w,\tau,\varepsilon) = \mathbf{F}^{(1)}(w,\tau,\varepsilon)(\mathbf{A}+\mathbf{B}+j\varepsilon w\mathbf{B}-x(\mathbf{I_0}-\mathbf{I_1}+j\varepsilon w\mathbf{I_1}))$$
$$+j\varepsilon\frac{\partial\mathbf{F}^{(1)}(w,\tau,\varepsilon)}{\partial w}(\mathbf{I_0}-\mathbf{I_1})+O(\varepsilon^2). \tag{18}$$

Solution of the problem (18) we can write in the form

$$\mathbf{F}^{(1)}(w,\tau,\varepsilon) = \Phi(w,\tau)\{\mathbf{R}+j\varepsilon w\mathbf{f}\}+O(\varepsilon^2), \tag{19}$$

where $\Phi(w,\tau)$ – scalar function whose form is defined below.
We obtain

$$j\varepsilon wa\Phi(w,\tau)\{\mathbf{R}+j\varepsilon w\mathbf{f}\} = \Phi(w,\tau)\{\mathbf{R}+j\varepsilon w\mathbf{f}\}(\mathbf{A}+\mathbf{B}+j\varepsilon w\mathbf{B}-x(\mathbf{I_0}-\mathbf{I_1}+j\varepsilon w\mathbf{I_1}))$$
$$+j\varepsilon\frac{\partial\Phi(w,\tau)}{\partial w}\{\mathbf{R}+j\varepsilon w\mathbf{f}\}+\Phi(w,\tau)j\varepsilon\mathbf{f}(\mathbf{I_0}-\mathbf{I_1})+O(\varepsilon^2),$$

then

$$j\varepsilon wa\Phi(w,\tau)\mathbf{R} = \Phi(w,\tau)\{\mathbf{R}(\mathbf{A}+\mathbf{B}-x(\mathbf{I_0}-\mathbf{I_1}$$
$$+j\varepsilon w[\mathbf{f}(\mathbf{A}+\mathbf{B}-x(\mathbf{I_0}-\mathbf{I_1}))+\mathbf{R}(\mathbf{B}-x\mathbf{I_1}]\}$$
$$+j\varepsilon\frac{\partial\Phi(w,\tau)}{\partial w}\{\mathbf{R}(\mathbf{I_0}-\mathbf{I_1})+o(\varepsilon^2).$$

Taking into account (8), divide the last equation by $j\varepsilon w$ and let $\varepsilon\to 0$, we get

$$a\mathbf{R} = \mathbf{f}(\mathbf{A}+\mathbf{B}-x(\mathbf{I_0}-\mathbf{I_1}))+\mathbf{R}(\mathbf{B}-x\mathbf{I_1})+\frac{\partial\Phi(w,\tau)/\partial w}{w\Phi(w,\tau)}\{\mathbf{R}(\mathbf{I_0}-\mathbf{I_1}),$$

we rewrite last equation

$$\mathbf{f}(\mathbf{A}+\mathbf{B}+x(\mathbf{I_1}-\mathbf{I_0})) = a\mathbf{R}-\mathbf{R}(\mathbf{B}-x\mathbf{I_1})+\frac{\partial\Phi(w,\tau)/\partial w}{w\Phi(w,\tau)}\{\mathbf{R}(\mathbf{I_1}-\mathbf{I_0}). \tag{20}$$

Solution $\mathbf{f}$ of the problem (20) we can write in the form

$$\mathbf{f} = C\mathbf{R}+\mathbf{g}-\varphi\frac{\partial\Phi(w,\tau)/\partial w}{w\Phi(w,\tau)}, \tag{21}$$

which we substitute to (20) and obtain

$$\varphi(\mathbf{A}+\mathbf{B}-x(\mathbf{I_0}-\mathbf{I_1})) = \mathbf{R}(\mathbf{I_0}-\mathbf{I_1}) \tag{22}$$

$$\mathbf{g}(\mathbf{A}+\mathbf{B}-x(\mathbf{I_0}-\mathbf{I_1})) = a\mathbf{R}+\mathbf{R}(x\mathbf{I_1}-\mathbf{B}). \tag{23}$$

Consider the first equation of system (8), differentiate it by $x$, we obtain the equation

$$\frac{\partial\mathbf{R}}{\partial x}\{(\mathbf{A}+\mathbf{B}-x(\mathbf{I_0}+\mathbf{I_1}))\}-\mathbf{R}(\mathbf{I_0}-\mathbf{I_1}) = 0.$$

Taking into account (22) and last equation for $\varphi$, we write important equality

$$\varphi = \frac{\partial \mathbf{R}}{\partial x}, \tag{24}$$

where $\varphi \mathbf{e} = 0$.

By virtue of (21), the vector $\mathbf{g}$ is a particular solution of the heterogeneous system (23), therefore, it satisfies some additional condition, which we will choose in the form $\mathbf{ge} = 0$, then the solution $\mathbf{g}$ of system (23), satisfying the condition $\mathbf{ge} = 0$, is determined uniquely. Now we consider the second equation of system (14), into which we substitute expansion (19)

$$\varepsilon^2 \frac{\partial \Phi(w,\tau)}{\partial \tau} + j\varepsilon w a \Phi(w,\tau)\{1 + j\varepsilon w \mathbf{fe}\}$$

$$= \left( j\varepsilon w + \frac{(j\varepsilon w)^2}{2} \right) \left( \Phi(w,\tau)\{\mathbf{R} + j\varepsilon w \mathbf{f}\}[\mathbf{B} - x\mathbf{I_0}] + j\varepsilon \frac{\partial \Phi(w,\tau)}{\partial w} \mathbf{R I_0} \right) \mathbf{e} + o(\varepsilon^3).$$

Then using Eq. (10)

$$\varepsilon^2 \frac{\partial \Phi(w,\tau)}{\partial \tau}$$

$$= \Phi(w,\tau) \left( (j\varepsilon w)^2 \{\mathbf{f}(\mathbf{B} - x\mathbf{I_0}) + x\mathbf{R I_0} - a\mathbf{f}\}\mathbf{e} + \frac{(j\varepsilon w)^2}{2} \mathbf{R}(\mathbf{B} - x\mathbf{I_0})\mathbf{e} \right)$$

$$+ (j\varepsilon)^2 w \frac{\partial \Phi(w,\tau)}{\partial w} \mathbf{R I_0} \mathbf{e},$$

we obtain following equation

$$\frac{\partial \Phi(w,\tau)/\partial \tau}{\partial \Phi(w,\tau)} = \frac{(jw)^2}{2} \{2(\mathbf{f}[\mathbf{B} - x\mathbf{I_0}] + \mathbf{R}x\mathbf{I_0} - a\mathbf{f})\mathbf{e} + a\}$$

$$- w \frac{\partial \Phi(w,\tau)/\partial w}{\partial \Phi(w,\tau)} \mathbf{R I_0} \mathbf{e},$$

to which we substitute solution (21)

$$\frac{\partial \Phi(w,\tau)/\partial \tau}{\partial \Phi(w,\tau)} = \frac{(jw)^2}{2} \{2\mathbf{g}[\mathbf{B} - x\mathbf{I_0}]\mathbf{e} + 2\mathbf{R}x\mathbf{I_0}\mathbf{e} + a\}$$

$$+ w \frac{\partial \Phi(w,\tau)/\partial w}{\partial \Phi(w,\tau)} \{\varphi[\mathbf{B} - x\mathbf{I_0}]\mathbf{e} - \mathbf{R I_0}\mathbf{e}\}. \tag{25}$$

Denoting

$$b(x) = a + 2\mathbf{g}[\mathbf{B} - x\mathbf{I_0}]\mathbf{e} + 2\mathbf{R}x\mathbf{I_0}\mathbf{e}, \tag{26}$$

we rewrite (25)

$$\frac{\partial \Phi(w,\tau)}{\partial \tau} = w \frac{\partial \Phi(w,\tau)}{\partial w)} \{\varphi[\mathbf{B} - x\mathbf{I_0}]\mathbf{e} - \mathbf{R I_0}\mathbf{e}\} + \frac{(jw)^2}{2} b(x) \Phi(w,\tau). \tag{27}$$

Let us consider

$$\varphi[\mathbf{B} - x\mathbf{I_0}]\mathbf{e} - \mathbf{R I_0}\mathbf{e},$$

using (24) to the last expression, we can get

$$\frac{\partial \mathbf{R}}{\partial x}[\mathbf{B} - x\mathbf{I_0}]\mathbf{e} - \mathbf{R I_0}\mathbf{e}. \tag{28}$$

Now consider Eq. (10) for $a(x)$, we differentiate $a(x)$ by $x$, and taking into account that the vector $\mathbf{R}$ as a solution of system (8) depends on $x$, we obtain

$$a'(x) = \frac{\partial \mathbf{R}}{\partial x}\mathbf{B}e - x\frac{\partial \mathbf{R}}{\partial x}\mathbf{I}_0 e - \mathbf{R}\mathbf{I}_0 e = \frac{\partial \mathbf{R}}{\partial x}(\mathbf{B} - x\mathbf{I}_0)e - \mathbf{R}\mathbf{I}_0 e.$$

Comparing this equality and (28), we rewrite Eq. (27) in the form

$$\frac{\partial \Phi(w,\tau)}{\partial \tau)} = a'(x)w\frac{\partial \Phi(w,\tau)}{\partial w} + b(x)\frac{(jw)^2}{2}\Phi(w,\tau). \tag{29}$$

Equation (30) is the Fourier transform of the Fokker-Planck equation for the probability distribution density $P(y,\tau)$ of the values of the centered and normalized number of calls in orbit. Find the inverse Fourier transform of expression (29), we obtain

$$\frac{\partial P(y,\tau)}{\partial \tau)} = -\frac{\partial}{\partial y}\{a'(x)yP(y,\tau)\} + \frac{1}{2}\frac{\partial^2}{\partial y^2}\{b(x)P(y,\tau)\}. \tag{30}$$

Since we compiled the Fokker-Planck equation for the function, therefore, this function is the probability density of the diffusion process, which we denote $y(\tau)$ with transport coefficient and diffusion coefficient $b(x)$. This process is a solution to the stochastic differential equation

$$dy(\tau) = a'(x)yd\tau + \sqrt{b(x)}dw(\tau). \tag{31}$$

Consider the stochastic process of the normalized number of calls in orbit

$$z(\tau) = x(\tau) + \varepsilon y(\tau), \tag{32}$$

where $\varepsilon = \sqrt{\sigma}$. We get (10), then $dx(\tau) = a(x)d\tau$, we obtain

$$dz(\tau) = d(x(\tau) + \varepsilon y(\tau)) = (a(x) + \varepsilon ya'(x))d\tau + \varepsilon\sqrt{b(x)}dw(\tau). \tag{33}$$

Consider following decompositions

$$a(z) = a(x + \varepsilon y) = a(x) + \varepsilon ya'(x) + O(\varepsilon^2),$$
$$\varepsilon\sqrt{b(z)} = \varepsilon\sqrt{b(x + \varepsilon y)} = \varepsilon\sqrt{b(x) + o(\varepsilon)} = \sqrt{b(x)} + O(\varepsilon^2).$$

Then we can rewrite Eq. (33) up to $O(\varepsilon^2)$ in the form:

$$dz(\tau) = a(z)d\tau + \sqrt{\sigma b(z)}dw(\tau). \tag{34}$$

We introduce the probability density function for the process $z(\tau)$

$$\pi(z,\tau) = \frac{\partial P\{z(\tau) < z\}}{\partial z}.$$

Since $z(\tau)$ is a solution of the stochastic differential Eq. (34), therefore, the process is a diffusion process and for its probability distribution density we can write the Fokker-Planck equation

$$\frac{\partial \pi(z, \tau)}{\partial \tau} = -\frac{\partial}{\partial z}\{a(z)\pi(z, \tau)\} + \frac{1}{2}\frac{\partial^2}{\partial z^2}\{\sigma b(z)\pi(z, \tau)\}.$$

We assume that the system operates in a stationary mode, then

$$\pi(z, \tau) = \pi(z),$$

therefore, the Fokker-Planck equation for the stationary probability distribution $\pi(z)$ has the form

$$(a(z)\pi(z))' + \frac{\sigma}{2}(b(z)\pi(z))'' = 0,$$

$$-a(z)\pi(z) + \frac{\sigma}{2}(b(z)\pi(z))' = 0.$$

Solving this differential equation, we obtain the probability distribution density $\pi(z)$ of the normalized number of calls in orbit in the form

$$\pi(z) = \frac{C}{b(z)}\exp\left\{\frac{\sigma}{2}\int_0^z \frac{a(x)}{b(x)}dx\right\}. \tag{35}$$

Using the found probability density $\pi(z)$, we compose a discrete probability distribution

$$P_1(i) = \pi(\sigma i)/\sum_{i=0}^{\infty} \pi(\sigma i), \tag{36}$$

which we will call the diffusion approximation of the probability distribution $P(i) = P(i(t) = i)$ of the number of calls in orbit in the $M|M|N$ system with feedback operating in the stationary mode.

It is not difficult to show that the condition for the existence of a stationary mode (ergodicity condition) of the system under consideration is the inequality

$$\lambda < r_0\mu N,$$

which we write in the form

$$\lambda = \beta r_0\mu N, \tag{37}$$

where $0 < \beta < 1$.

For any approximation, including for (36), it is fundamentally important to determine its accuracy and area of applicability, i.e. the range of those values of the network load parameters $\beta$ and parameter $\sigma$, the values of which in theoretical studies are infinitesimal ($\sigma \to 0$).

The accuracy of the approximation will be determined by the Kolmogorov distance

$$\Delta = \max_{0 \le i < \infty} \left| \sum_{n=0}^{i} \left(P(n) - P_1(n)\right) \right|. \tag{38}$$

Naturally, to find the values of $\Delta$, it is necessary to know the initial probability distribution $P(i)$. We will find this probability distribution in the particular case for the feedback system under consideration for $N = 1$, i.e. for the single-line feedback system $M|M|N$.

It can be shown that for a single-line system, the characteristic function

$$h(u) = \sum_{i=0}^{\infty} e^{jui} P(i)$$

has the form

$$h(u) = (1 + \alpha - \alpha e^{ju}) \left( \frac{1 - \alpha}{1 - \alpha e^{ju}} \right)^s,$$

where

$$\alpha = \frac{\lambda}{r_0 \mu}, \quad s = \frac{\lambda + r_2 \mu + \sigma}{\sigma}.$$

The probability distribution $P(i)$ can be written as

$$P(i) = \frac{1}{2\pi} \int_{-\pi}^{\pi} e^{-jui} h(u) du.$$

For $\mu = 1$, $r_0 = 0.5$, $r_1 = 0.3$, $r_2 = 0.2$, the table below shows the $\Delta$ values from (39) and the indicated values of the parameters $\beta$ and $\sigma$ (Table 1).

**Table 1.** Kolmogorov distance

| | $\sigma = 5$ | $\sigma = 1$ | $\sigma = 0.5$ | $\sigma = 0.2$ | $\sigma = 0.1$ | $\sigma = 0.05$ |
|---|---|---|---|---|---|---|
| $\beta = 0.5$ | 0.053 | 0.008 | 0.039 | 0.043 | 0.029 | 0.012 |
| $\beta = 0.6$ | 0.044 | 0.026 | 0.042 | 0.033 | 0.022 | 0.015 |
| $\beta = 0.7$ | 0.029 | 0.032 | 0.036 | 0.023 | 0.016 | 0.011 |
| $\beta = 0.8$ | 0.013 | 0.027 | 0.024 | 0.014 | 0.010 | 0.007 |
| $\beta = 0.9$ | 0.002 | 0.015 | 0.011 | 0.007 | 0.005 | 0.003 |

Assuming that the approximation $P_1(i)$ is acceptable if its accuracy is $\Delta < 0.05$, we can conclude.

The proposed diffusion approximation $P_1(i)$ is acceptable in almost the entire spectrum of values of the parameters $\beta$ and $\sigma$. The accuracy of the approximation increases ($\Delta$ decreases) with decreasing values of the $\sigma$. This is quite natural due to the limiting condition $\sigma \to 0$.

An unobvious result is that the proposed approximation $P_1(i)$ is also acceptable for sufficiently large values of $0.2 < \sigma \le 5$.

As the load increases with $\beta \ge 0.6$, the accuracy of the approximation $P_1(i)$ also increases, while $\Delta$ reaches values less than 0.01, which indicates a very high accuracy of the proposed method.

# 5    Conclusion

Thus, a mathematical model of $M|M|N$ type queuing system was constructed, probability distribution of number of the customers in the orbit was found by unique method of asymptotic diffusion analysis, which allows to increase the accuracy of approximation, as well as expand the area of applicability up to 10 times.

# References

1. Avrachenkov, K., Yechiali, U.: Retrial networks with finite buffers and their application to internet data traffic. Probab. Eng. Inf. Sci. **22**, 519–536 (2008)
2. Roszik, J., Sztrik, J., Kim, C.S.: Retrial queues in the performance modelling of cellular mobile networks using MOSEL. Int. J. Simul. **6**, 38–47 (2005)
3. Phung-Duc, T.: Retrial Queueing Models: A Survey on Theory and Applications. arXiv 2019, arXiv:1906.09560v1
4. Shekhar, C., Raina, A.A., Kumar, A.: A brief review on retrial queue: progress in 2010–2015. Int. J. Appl. Sci. Eng. Res. **5**, 324–336 (2016)
5. Kim, C., Mushko, V., Dudin, A.: Computation of the steady state distribution for multi-server retrial queues with phase type service process. Ann. Oper. Res. **201**, 307–323 (2012)
6. Phung-Duc, T., Masuyama, H., Kasahara, S., Takahashi, Y.: A matrix continued fraction approach to multiserver retrial queues. Ann. Oper. Res. **203**, 61–183 (2013)
7. Horling, B., Lesser, V.: Using queueing theory to predict organizational metrics. In: AAMAS'06, Hakodate, Japan, pp. 1098–1100 (2006)
8. Gnanasambandam, N., Lee, S.C., Gautam, N., et al.: Reliable MAS performance prediction using queueing models. In: IEEE First Symposium on Multi-Agent Security and Survivalibility, pp. 55–64 (2004)
9. Gnanasmbandam, N., Lee, S., Kumara, S.R.T.: An autonomous performance controlframework for distributed multi-agent systems: a queueing theory based approach. In: AAMAS'05, Utrecht, Netherlands, 25–29 July 2005, pp. 1313–1314
10. Lee, M.H., Birukou, A., Dudin, A., Klimenok, V., Choe, C.: Quantitative analysis of single-level single-mediator multi-agent systems. In: Nguyen, N.T., Grzech, A., Howlett, R.J., Jain, L.C. (eds.) KES-AMSTA 2007. LNCS (LNAI), vol. 4496, pp. 447–455. Springer, Heidelberg (2007). https://doi.org/10.1007/978-3-540-72830-6_46
11. Moiseeva, E., Nazarov, A.: Asymptotic analysis of RQ-systems M/M/1 on heavy load condition. In: Proceedings of the IV International Conference Problems of Cybernetics and Informatics, Baku, Azerbaijan, 12–14 September 2012, pp. 164–166
12. Fedorova, E.: Quasi-geometric and gamma approximation for retrial queueing systems. In: Dudin, A., Nazarov, A., Yakupov, R., Gortsev, A. (eds.) ITMM 2014. CCIS, vol. 487, pp. 123–136. Springer, Cham (2014). https://doi.org/10.1007/978-3-319-13671-4_15
13. Nazarov, A., Chernikova, Y.: Gaussian approximation of distribution of states of the retrial queueing aystem with r-persistent exclusion of alternative customers. In: Dudin, A., Nazarov, A., Yakupov, R. (eds.) ITMM 2015. CCIS, vol. 564, pp. 200–208. Springer, Cham (2015). https://doi.org/10.1007/978-3-319-25861-4_17

14. Lapatin, I., Nazarov, A.: Asymptotic analysis of the output process in retrial queue with markov-modulated poisson input under low rate of retrials condition. In: Vishnevskiy, V.M., Samouylov, K.E., Kozyrev, D.V. (eds.) DCCN 2019. CCIS, vol. 1141, pp. 315–324. Springer, Cham (2019). https://doi.org/10.1007/978-3-030-36625-4_25
15. Falin, G.I., Templeton, J.G.C.: Retrial Queues. Chapman & Hall, London (1997)
16. Falin, G.I.: A survey of retrial queues. Queueing Syst. **7**, 127–168 (1990)
17. Artalejo, J.R., Gomez-Corral, A.: Retrial Queueing Systems: A Computational Approach. Springer, Berlin/Heidelberg (2008)
18. Nazarov, A., Phung-Duc, T., Paul, S., Lizura, O.: Single server queues with batch poisson input and multiple types of outgoing calls. In: Dudin, A., Nazarov, A., Moiseev, A. (eds.) ITMM 2019. CCIS, vol. 1109, pp. 177–187. Springer, Cham (2019). https://doi.org/10.1007/978-3-030-33388-1_15
19. Moiseev, A., Nazarov, A., Paul, S.: Asymptotic diffusion analysis of multi-server retrial queue with hyper-exponential service. Mathematics 8, 531 (2020)
20. Nazarov, A., Phung-Duc, T., Paul, S.: Slow retrial asymptotics for a single server queue with two-way communication and markov modulated poisson input. J. Systems Science and Syst. Eng. **28**(2), 81–193 (2019)

# Exact Performance Analysis of Retrial Queues with Collisions

Tuan Phung-Duc[1][✉] and Dieter Fiems[2]

[1] University of Tsukuba, Tsukuba, Japan
`tuan@sk.tsukuba.ac.jp`
[2] Ghent University, Ghent, Belgium
`Dieter.Fiems@UGent.be`

**Abstract.** This paper considers the performance of a collision channel where the transmissions of packets are split in two phases. Arriving packets who see the transmission channel idle, occupy it immediately. Arriving packets cannot detect ongoing transmissions during the first phase of transmission. Hence, if the first part or phase of the packet is being transmitted, a collision occurs and both packets join a virtual orbit and retry to access the channel later on. In case an arriving packet sees the transmission in the second phase, the arriving packet is blocked and joins the orbit, while the transmission of the other packet continues uninterrupted. From the orbit, packets independently retry to occupy the channel after an exponentially distributed time. The behavior of retrial packets and fresh ones are identical. Assuming Poisson arrivals and exponentially distributed packet lengths (for each phase), we obtain an exact expression for the joint generating function of the number of packets in the orbit and the state of the transmission channel. We then study how the division into phases affects queueing performance.

## 1 Introduction

This paper is motivated by collision detection protocols for wireless networks. In wireless networks, multiple nodes share a channel for transmitting data. Therefore, it is possible that one user starts sending while another one is occupying the channel. Multiple access protocols are therefore required to regulate how users access the transmission channels, one of the most famous protocols being the CSMA/CD (Carrier Sense Multiple Access with Collision Detection) protocol in which nodes use their sensing capability to avoid collisions. Although nodes have sensing capability, collisions are not completely avoidable. Once a collision occurs, both the node in transmission and the colliding one must retransmit later on.

Once a user has data to transmit, it senses to know whether or not the channel is idle. If the channel is idle, the node transmits its data. Otherwise it senses again after a random time. However, since the sensing capability is not perfect, transmissions of the nodes may collide with ongoing transmissions over the channel. This is in particular the case if another transmission just

© Springer Nature Switzerland AG 2020
M. Gribaudo et al. (Eds.): ASMTA 2019, LNCS 12023, pp. 144–157, 2020.
https://doi.org/10.1007/978-3-030-62885-7_11

started. In case of a collision, both users try to retransmit after a random time. In practice, if the ongoing node has been transmitting for a long enough time, other users are able to sense its presence. As a result, there is no collision and other users will retry after some random time. Motivated by this situation, we consider a retrial queue with single server queue with two phases of service. In the first phase, collisions may occur while in the second phase there is no collision at all.

This model has been treated in [1] under a general setting with arbitrary service distribution of the second phase. In this paper, we revisit this model and consider the case where the phases of the service times are exponentially distributed. This allows for obtaining more explicit and detailed expressions for the generating function of the orbit lengths. We also consider an extension where fresh customers may abandon the system upon collision or blocking. This apparently slight complication considerably affects the performance analysis.

Also [2,3] investigate retrial queues with collisions. In [2], similar to the model at hand, the author considers a retrial queue in which the transmission has a collision phase followed by a non-collision phase. The collision phase has a fixed length while the non-collision phase follows an arbitrary distribution. The main feature of the model in [2] is that the retrial mechanism is the so-called constant retrial policy in which the retrial rate does not depend on the number of retrial customers. This assumption comes natural if every user is able to detect the status of the other users. Kumar et al. [3] consider a single collision channel with only the collision phase and feedback to customers in the Markovian setting. Furthermore, Kumar et al. [4] investigate the retrial queue with feedback for the constant retrial policy, assuming arbitrarily distributed service and retrial time distributions.

The remainder of the paper is organized as follows. In Sect. 2, we describe the model in detail, while Sect. 3 presents its performance analysis. Section 4 presents an extension of the model while some numerical results are presented in Sect. 5. Finally, concluding remarks are given in Sect. 6.

## 2  Model

We consider a single channel in a random access network. Packets arrive at the channel according to a Poisson process with fixed rate $\lambda$. If the channel is idle, the packet can occupy the channel immediately. The channel holding time of a packet consists of two phases: in the first phase other users cannot detect the transmission of the packet and therefore it is possible that the transmission of the packet collides with the transmission of other packets. If this is the case, both packets have to retry accessing the channel later on. After completing the first phase without collisions, the packet moves to the second phase during which no more collisions occur. After the second phase, the packet will depart from the channel. We assume that the duration of the first and second phases follow exponential distributions with rates $\alpha$ and $\mu$, respectively. Moreover, the arrival process and the consecutive lengths of the phases are assumed to be independent random processes.

When there is a collision both packets join a virtual orbit. Packets also join the orbit when they arrive while the second phase of a transmission is going on. Packets remain in the orbit for a random amount of time, the retrial times constituting a sequence of independent exponentially distributed random variables with common retrial rate $\nu$. Whenever packets come out of orbit, they are treated as if they were new packets.

## 3    Analysis

In view of the assumptions above, the state of the queueing system at hand is described by the tuple $(C(t), N(t))$, where $N(t)$ denotes the number of packets in orbit and where $C(t)$ denotes the state of the channel, which is defined as follows:

$$C(t) = \begin{cases} 0 & \text{if the channel is free,} \\ 1 & \text{if the first phase of a packet transmission is ongoing,} \\ 2 & \text{if the second phase of a packet transmission is ongoing.} \end{cases}$$

The state space of this Markov chain is $\{0,1,2\} \times Z_+$, where $Z_+ = 0,1,2,\ldots$ denotes the set of non-negative integers.

Assuming the stationary distribution exists, let $\pi_{i,j}$ denote the stationary probability,

$$\pi_{i,j} = \lim_{t \to \infty} P[C(t) = i, N(t) = j].$$

For $C(t) = 0$, a new arrival (with rate $\lambda$) induces a transition to state $(1, N(t))$, while a retrial (with rate $\nu N(t)$) induces a transition to state $(1, N(t) - 1)$ as a packet in orbit can occupy the channel. For $C(t) = 1$, a new arrival (with rate $\lambda$) induces a transition to state $(0, N(t) + 2)$ as both the packet being transmitted and the new packet join the orbit. A retrial (with rate $\nu N(t)$) induces a transition to state $(0, N(t) + 1)$ as the packet coming out of orbit causes a collision. Finally, with rate $\alpha$, the packet finishes phase 1, the new state being $(2, N(t))$. For $C(t) = 2$, a new arrival (with rate $\lambda$) directly moves to the orbit, the new state being $(2, N(t) + 1)$, while with rate $\mu$ the packet transmission ends, inducing a transition to state $(0, N(t))$. In view of the possible transitions from the different states, the balance equations read,

$$\begin{aligned} (\lambda + j\nu)\pi_{0,j} &= \mu\pi_{2,j} + \lambda\pi_{1,j-2} + (j-1)\nu\pi_{1,j-1}, \\ (\lambda + j\nu + \alpha)\pi_{1,j} &= \lambda\pi_{0,j} + (j+1)\nu\pi_{0,j+1}, \\ (\lambda + \mu)\pi_{2,j} &= \alpha\pi_{1,j} + \lambda\pi_{2,j-1}, \end{aligned} \tag{1}$$

where we set $\pi_{i,-1} = \pi_{i,-2} = 0$ for ease of notation.

We now introduce the following partial probability generating functions,

$$\Pi_i(z) = \sum_{j=0}^{\infty} \pi_{i,j} z^j, \qquad |z| \le 1. \tag{2}$$

By standard $z$-transform techniques, we transform the balance equations into the following set of differential equations for the corresponding generating functions,

$$\lambda \Pi_0(z) + \nu z \Pi_0'(z) = \mu \Pi_2(z) + \lambda z^2 \Pi_1(z) + \nu z^2 \Pi_1'(z), \tag{3}$$

$$(\lambda + a)\Pi_1(z) + \nu z \Pi_1'(z) = \lambda \Pi_0(z) + \nu \Pi_0'(z), \tag{4}$$

$$(\lambda + \mu)\Pi_2(z) = a\Pi_1(z) + \lambda z \Pi_2(z). \tag{5}$$

To solve this set of equations, we first note that summing the three equations above and rearranging the result leads to

$$\nu \Pi_0'(z) = \nu z \Pi_1'(z) + \lambda(z+1)\Pi_1(z) + \lambda \Pi_2(z). \tag{6}$$

The left hand side of this equation represents the flow going out the orbit while the right hand side represents the flows going into the orbit. It follows from (4) and (6) that

$$\lambda(z+1)\Pi_1(z) + \lambda \Pi_2(z) = (\lambda + a)\Pi_1(z) - \lambda \Pi_0(z),$$

leading to

$$(\lambda z - a)\Pi_1(z) + \lambda \Pi_2(z) = -\lambda \Pi_0(z).$$

Moreover, solving (5) for $\Pi_2(z)$ yields,

$$\Pi_2(z) = \frac{a\Pi_1(z)}{\lambda + \mu - \lambda z},$$

By the expressions above, we can therefore express both $\Pi_1(z)$ and $\Pi_2(z)$ in terms of $\Pi_0(z)$. We have,

$$\Pi_1(z) = \frac{\lambda(\lambda + \mu - \lambda z)}{\lambda^2 z^2 - \lambda z(\lambda + \mu + a) + \mu a} \Pi_0(z), \tag{7}$$

$$\Pi_2(z) = \frac{\lambda a}{\lambda^2 z^2 - \lambda z(\lambda + \mu + a) + \mu a} \Pi_0(z). \tag{8}$$

In particular, evaluating these expressions for $z = 1$ yields,

$$\Pi_1(1) = \frac{\lambda \mu}{\mu a - \lambda(\mu + a)} \Pi_0(1),$$

$$\Pi_2(1) = \frac{\lambda a}{\mu a - \lambda(\mu + a)} \Pi_0(1).$$

Moreover, the normalization condition of generating functions requires $\Pi_0(1) + \Pi_1(1) + \Pi_2(1) = 1$. Hence, we can solve $\Pi_i(1)$, $i = 1, 2, 3$. After some simplifications, we find the following expressions.

**Theorem 1.** *The distribution of the state of the channel is given by the following expressions:*

$$\Pi_0(1) = 1 - \lambda\left(\frac{1}{\mu} + \frac{1}{a}\right), \quad \Pi_1(1) = \frac{\lambda}{a}, \quad \Pi_2(1) = \frac{\lambda}{\mu}.$$

To simplify notation, let $\alpha(z)$ denote the following rational function,

$$\alpha(z) = \frac{\lambda(\lambda + \mu - \lambda z)}{\lambda^2 z^2 - \lambda z(\lambda + \mu + \alpha) + \mu\alpha} \, ,$$

such that we have $\Pi_1(z) = \alpha(z)\Pi_0(z)$. Substituting this expression into (4), we obtain

$$((\lambda + \alpha)\alpha(z) + \nu z\alpha'(z) - \lambda)\Pi_0(z) = \nu(1 - z\alpha(z))\Pi_0'(z) \, ,$$

or, equivalently,

$$\Pi_0'(z) = \gamma(z)\Pi_0(z) \, , \tag{9}$$

with,

$$\gamma(z) = \frac{(\lambda + \alpha)\alpha(z) + \nu z\alpha'(z) - \lambda}{\nu(1 - z\alpha(z))} \, .$$

Equation (9) is a first-order differential equation, and therefore easily solved, accounting for the value $\Pi_0(1)$ as given in Theorem 1. Summarizing, we have the following result.

**Theorem 2.** $\Pi_0(z)$ is given by

$$\Pi_0(z) = \Pi_0(1) \exp\left(-\int_z^1 \gamma(u)du\right) \, , \tag{10}$$

for $|z| \leq 1$, while $\Pi_1(z)$ and $\Pi_2(z)$ are given by (7) and (8).

So far, we have assumed that the stationary distribution exists. We now investigate the necessary and sufficient stability conditions. First note that after some calculations $\gamma(1)$ can be expressed as follows,

$$\gamma(1) = \frac{\lambda^2[(2\mu + \alpha)(\mu\alpha - \lambda(\mu + \alpha)) + \nu(\mu^2 + \lambda\alpha)]}{\nu(\mu\alpha - \lambda(\mu + \alpha))(\mu\alpha - \lambda(2\mu + \alpha))} \, . \tag{11}$$

As we need $\gamma(1) > 0$, see (9), the following inequality is a necessary condition for the existence of the stationary distribution,

$$\lambda\left(\frac{2}{\alpha} + \frac{1}{\mu}\right) < 1 \, . \tag{12}$$

We now also show that this is a sufficient condition. To this end, we study the Markov process at epochs when the server enters state 0. Let $\tau_n$ denote the $n$th such epoch and let $N_n \doteq N(\tau_n)$ denote the size of the orbit at these epochs. Obviously, the process $N_n$ is a discrete-time Markov process. We will now show that this discrete-time Markov process satisfies the following drift-condition,

$$\mathsf{E}[N_{n+1} - m|N_n = m] \leq 0 \text{ for } m > M \, ,$$

for some constant $M$. This negative drift condition implies ergodicity of the discrete-time Markov process [5]. Moreover, as $\mathsf{E}[\tau_{n+1} - \tau_n] < 1/\lambda + 1/\alpha + 1/\mu$ is finite, the ergodicity of $N_n$ implies the ergodicity of the original Markov process. First consider the case when there are no collisions in $[\tau_n, \tau_{n+1})$. There are two possibilities. If phase 1 starts with a new arrival, all arrivals during phase 2 join the orbit (there are $\lambda/\mu$ such arrivals on average). On the other hand, if phase 1 starts with a retrial, the orbit also contains all arrivals during phase 2, but no longer contains the retrying customer. This results in a drift,

$$d_0(m) = E[(N_{n+1} - m)1_{\{\text{no collisions in } [\tau_n, \tau_{n+1})\}} | N_n = m]$$

$$= \frac{\lambda}{\lambda + \nu m} \frac{\alpha}{\alpha + \lambda + m\nu} \frac{\lambda}{\mu} + \frac{\nu m}{\lambda + \nu m} \frac{\alpha}{\alpha + \lambda + (m-1)\nu} \left(\frac{\lambda}{\mu} - 1\right),$$

the second factor in each of the terms being the probability that there are no collisions during phase 1. ($1_{\{.\}}$ is the indicator function which evaluates to 1 if its argument is true and to 0 if this is not the case.)

Let $d_1(n)$ denote the drift when there are collisions in phase 1. We have the following scenarios: (i) phase 1 starts with a new arrival and another new arrival collides; (ii) phase 1 starts with a new arrival and a retrial collides; (iii) phase 1 starts with a retrial and a new arrival collides; and (iv) phase 1 starts with a retrial and a retrial collides. Noting that the orbit size only increases with every new arrival, we find the following drift,

$$d_1(m) = E[(N_{n+1} - m)1_{\{\text{collision in } [\tau_n, \tau_{n+1})\}} | N_n = m]$$

$$= 2\frac{\lambda}{\lambda + \nu m} \frac{\lambda}{\lambda + m\nu + \alpha} + \frac{\lambda}{\lambda + \nu m} \frac{m\nu}{\lambda + m\nu + \alpha} + \frac{m\nu}{\lambda + m\nu} \frac{\lambda}{\lambda + m\nu + \alpha}$$

$$= \frac{2\lambda}{\lambda + m\nu + \alpha}.$$

Combining these expressions, we further find after some manipulations,

$$E[N_{n+1} - m | N_n = m] = d_0(m) + d_1(m) = \xi_1(m) \left(\frac{\lambda}{\mu} + 2\frac{\lambda}{\alpha} - 1 + \xi_2(m)\right),$$

with,

$$\xi_1(m) = \frac{\alpha \left(m^2\nu^2 + \nu m\alpha + \lambda\nu(2m-1) + \lambda\alpha + \lambda^2\right)}{(m\nu + \alpha + \lambda)(m\nu + \lambda)((m-1)\nu + \alpha + \lambda)},$$

$$\xi_2(m) = \frac{\lambda \left(\nu m\alpha - 2m\nu^2 + \alpha^2 + \lambda\alpha - \alpha\nu\right)}{\alpha \left(m^2\nu^2 + \nu m\alpha + 2\lambda m\nu + \lambda\alpha + \lambda^2 - \lambda\nu\right)}.$$

It is now easy to verify that $\xi_1(m) > 0$ for $m \geq 1$. Moreover, $\lim_{m \to \infty} \xi_2(m) = 0$, hence there exists a $M$ such that $\lambda/\mu + 2\lambda/\alpha - 1 + \xi_2(m) < 0$ for all $m > M$. We conclude that the drift is non-positive for $m > M$.

**Theorem 3.** *The necessary and sufficient stability condition for our model is (12).*

We now investigate the tail asymptotics of the joint stationary distributions using the method presented in [6]. To this end, we investigate the poles of $\gamma(z)$. By some trivial calculations, we find the following four poles,

$$z_1 = \frac{\lambda + \mu + \alpha - \sqrt{(\lambda + \mu + \alpha)^2 - 4\mu\alpha}}{2\lambda},$$

$$z_2 = \frac{\lambda + \mu + \alpha + \sqrt{(\lambda + \mu + \alpha)^2 - 4\mu\alpha}}{2\lambda},$$

$$z_3 = \frac{2(\lambda + \mu) + \alpha - \sqrt{(2\mu + \alpha + 2\lambda)^2 - 8\mu\alpha}}{4\lambda},$$

$$z_4 = \frac{2(\lambda + \mu) + \alpha + \sqrt{(2\mu + \alpha + 2\lambda)^2 - 8\mu\alpha}}{4\lambda}.$$

One shows that $z_3 < z_1$, while we also have $z_3 < z_4$ and $z_1 < z_2$. Therefore $z_3$ is the dominant pole of $\gamma(z)$. As such, we can write $\gamma(z)$ in the following form,

$$\gamma(z) = \Phi(z) + \frac{a}{z - z_3}, \tag{13}$$

where $\Phi(z)$ is an analytic function in $z < z_3 + \epsilon$, with $\epsilon$ a small enough positive number, and with $a$ the residue of the pole $z_3$,

$$a = \lim_{z \to z_3} \gamma(z)(z - z_3).$$

In view of (10) and (13), we find that $\Pi_0(z)$ can be rewritten as follows,

$$\Pi_0(z) = \Pi_0(1) \exp\left(-\int_z^1 \Phi(u)du\right) \left(\frac{z_3 - 1}{z_3 - z}\right)^{-a}$$

which in turn leads to the following asymptotic approximation of $\pi_{0,j}$.

$$\pi_{0,j} \asymp c_0 j^{-a-1} z_3^{-j}, \qquad j \to \infty,$$

where $c_0$ is an explicit constant. A similar investigation of the singularities of $\Pi_1(z)$ and $\Pi_2(z)$ shows that

$$\pi_{1,j} \asymp c_1 j^{-a-1} z_3^{-j}, \qquad j \to \infty,$$
$$\pi_{2,j} \asymp c_2 j^{-a-1} z_3^{-j}, \qquad j \to \infty.$$

We summarize these results in the following theorem.

**Theorem 4.** *We have the following asymptotic formulae for the joint stationary distribution.*

$$\pi_{i,j} \asymp c_i j^{a-1} z_3^{-j}, \qquad j \to \infty, \tag{14}$$

*where $c_i$ $(i = 0, 1, 2)$ are given in explicit form.*

*Remark 1.* It is worth to note that the order of $\pi_{i,j}$ is the same for $i = 0, 1, 2$. This is different from conventional retrial queues in which $\pi_{1,j}$ typically has a larger order than $\pi_{0,j}$.

We can also calculate the different moments of the orbit size by evaluating the derivatives of the probability generating functions $\Pi_i(z)$ for $z = 1$. We have the following results.

**Theorem 5.** *The mean number of customers in the orbit are given by the following expressions.*

$$\Pi_0'(1) = \frac{\lambda^2}{\mu \nu} + \frac{\mu (2\alpha - 2\lambda + \nu)\lambda^2}{(\alpha(\mu - \lambda) - 2\lambda\mu)\nu\alpha} - \frac{(\mu - \nu)\lambda^3}{(\alpha(\mu - \lambda) - 2\lambda\mu)\mu\nu},$$

$$\Pi_1'(1) = \frac{(\lambda\mu\alpha + \alpha\lambda\nu + 2\lambda\mu^2 + \mu^2\nu)\lambda^2}{(\alpha(\mu - \lambda) - 2\lambda\mu)\nu\mu\alpha},$$

$$\Pi_2'(1) = \frac{(\alpha\lambda + \alpha\nu + 2\lambda\mu - 2\nu\lambda + \mu\nu)\lambda^2}{(\alpha(\mu - \lambda) - 2\lambda\mu)\nu\mu}.$$

*Proof.* The first equation follows from plugging $z = 1$ and (11) in (9). The other two equalities are obtained by substituting $z = 1$, the expression for $\Pi_0'(1)$ and the expressions in Theorem 1 into the differential equations (4) and (5).

The expressions above also allow for determining the mean number $\overline{N}$ of customers in orbit. After some simplifications, we find,

$$\overline{N} = \sum_{i=0}^{2} \Pi_i'(1) = \frac{(\mu + \nu)\alpha^2 + \mu (2\mu + \nu)\alpha + 2\mu^2\nu}{((\mu - \lambda)\alpha - 2\lambda\mu)\nu\mu\alpha}\lambda^2.$$

As we have found explicit expressions for the generating functions $\Pi_i(z)$, we can in principle extract the joint stationary distribution by inverting these generating functions. Such an inversion approach is however fairly complex in comparison to the direct calculations below. Indeed, the following recursive scheme can be used to calculate the stationary joint orbit length distribution.

**Theorem 6.** *The stationary joint orbit length distribution can be recursively computed as follows:*

$$\pi_{0,j+1} = \frac{(\lambda + j\nu + \alpha)\pi_{1,j} - \lambda\pi_{0,j}}{(j + 1)\nu},$$

$$\pi_{2,j+1} = \frac{(\lambda + (j + 1)\nu)\pi_{0,j+1} - \lambda\pi_{1,j-1} - j\nu\pi_{1,j}}{\mu},$$

$$\pi_{1,j+1} = \frac{(\lambda + \mu)\pi_{2,j+1} - \lambda\pi_{2,j}}{\alpha},$$

*with,*

$$\pi_{0,0} = \Pi_0(1)\exp\left(-\int_0^1 \gamma(u)du\right), \quad \pi_{1,0} = \frac{\lambda(\lambda + \mu)}{\mu\alpha}\pi_{0,0}, \quad \pi_{2,0} = \frac{\lambda}{\mu}\pi_{0,0}.$$

*Proof.* The initial value $\pi_{0,0} = \Pi_0(0)$ is obtained by evaluating (10) in $z = 0$. Moreover, the expressions for $\pi_{1,0} = \Pi_1(0)$ and $\pi_{2,0} = \Pi_2(0)$ follow from evaluating (7) and (8) in $z = 0$. Finally, the recursive expressions for higher orbit sizes directly follow from the balance equations (1).

*Remark 2.* Note that $\gamma(z)$ is a rational function of $z$ with known poles ($z_i$, $i = 1, 2, 3, 4$). Therefore, the integral in Theorem 6 can be obtained explicitly.

*Remark 3.* The recursive calculations can be combined with the asymptotic approximation. More precisely, it suffices to calculate the first 100 or so terms by the recursion, terms for higher orbit sizes being well approximated by the asymptotic approximations.

## 4   Extension to Models with Abandonment

We now consider the following extension. If a newly arriving customer collides with the customer in service, the arriving customer will join the orbit with probability $p$ or abandon with probability $\bar{p} = 1 - p$. In case, a retrial customer collides with the customer in service, the retrial customer will always join the orbit, as well as a customer in service that collide. Assuming that the Markov process at hand is stable, we now have the following set of balance equations,

$$(\lambda + j\nu)\pi_{0,j} = \mu\pi_{2,j} + \lambda p\pi_{1,j-2} + \lambda\bar{p}\pi_{1,j-1} + (j-1)\nu\pi_{1,j-1}$$
$$(\lambda + j\nu + \alpha)\pi_{1,j} = \lambda\pi_{0,j} + (j+1)\nu\pi_{0,j+1},$$
$$(\lambda + \mu)\pi_{2,j} = \alpha\pi_{1,j} + \lambda\pi_{2,j-1},$$

where we set $\pi_{i,-1} = \pi_{i,-2} = 0$ as before to simplify notation.

Transforming these equations to those for the generating function yields,

$$\lambda\Pi_0(z) + \nu z\Pi_0'(z) = \mu\Pi_2(z) + \lambda(pz + \bar{p})z\Pi_1(z) + \nu z^2\Pi_1'(z),$$
$$(\lambda + \alpha)\Pi_1(z) + \nu z\Pi_1'(z) = \lambda\Pi_0(z) + \nu\Pi_0'(z),$$
$$(\lambda + \mu)\Pi_2(z) = \alpha\Pi_1(z) + \lambda z\Pi_2(z).$$

As always, summing up these equations and dividing both sides by $z - 1$ we obtain the following orbit balancing equation,

$$(\lambda pz + \lambda)\Pi_1(z) + \nu z\Pi_1'(z) + \lambda\Pi_2(z) = \nu\Pi_0'(z)$$

Combining this equation and the second one in the system of equations above we obtain

$$\Pi_1(z) = \frac{\lambda(\lambda + \mu - \lambda z)}{\lambda^2 pz^2 - \lambda z(\lambda p + \mu p + \alpha) + \mu\alpha}\Pi_0(z),$$

$$\Pi_2(z) = \frac{\lambda\alpha}{\lambda^2 pz^2 - \lambda z(\lambda p + \mu p + \alpha) + \mu\alpha}\Pi_0(z).$$

By evaluating the former expressions in $z = 1$ and by using the normalization condition $\Pi_0(1) + \Pi_1(1) + \Pi_2(1) = 1$, we obtain the distribution of the state of the server.

**Theorem 7.** *The probabilities that server is in the different states are given by the following expressions:*

$$\Pi_0(1) = \frac{\mu\alpha - \lambda(\mu p + \alpha)}{\mu(\lambda + \alpha) - \lambda\mu p}, \quad \Pi_1(1) = \frac{\lambda\mu}{\mu(\lambda + \alpha) - \lambda\mu p},$$

$$\Pi_2(1) = \frac{\lambda\alpha}{\mu(\lambda + \alpha) - \lambda\mu p}.$$

As in the previous section, let the functions $\alpha(z)$ and $\gamma(z)$ be defined such that $\Pi_1(z) = \alpha(z)\Pi_0(z)$ and $\Pi_0(z)' = \gamma(z)\Pi_0(z)$. After some calculations, we find,

$$\gamma(z) = \frac{(\lambda + \alpha)\alpha(z) + \nu z\alpha'(z) - \lambda}{\nu(1 - z\alpha(z))},$$

and,

$$\gamma(1) = \frac{\lambda^2\left[(\mu\alpha - \lambda(\mu p + \alpha))(\mu(1 + p) + \alpha) + \nu(\mu^2 p + \lambda\alpha)\right]}{\nu(\mu\alpha - \lambda(\mu p + \alpha))(\mu\alpha - \lambda\mu(1 + p) - \lambda\alpha)}.$$

From $\gamma(1) > 0$, we find that a necessary assumption for stability is

$$\lambda\left(\frac{1 + p}{\alpha} + \frac{1}{\mu}\right) < 1.$$

By studying the drift of the Markov process when the server is idle, we can again show that this stability condition is sufficient to guarantee a negative drift. This is sufficient to conclude that the restricted Markov process is ergodic from which the ergodicity of the unrestricted chain follows.

Finally, we can again calculate the various moments of the orbit size. Using the moment-generating property of probability generating functions, we obtain the following expressions after some calculations.

**Theorem 8.** *The mean number of customers in the orbit are given by the following expressions.*

$$\Pi_0'(1) = \frac{2\alpha\lambda\mu p + \mu^2 p(\lambda p - \alpha + \lambda - \nu) + \alpha^2(\lambda - \mu) + \alpha\lambda(\mu - \nu) - \alpha\mu^2}{(\alpha + \lambda\bar{p})\mu\nu(\lambda\mu p + \alpha\lambda - \alpha\mu + \lambda\mu)}\lambda^2,$$

$$\Pi_1'(1) = \frac{\lambda\mu^2(p + 1) + \mu^2\nu p + \mu\alpha\lambda + \alpha\lambda\nu}{(\alpha + \lambda\bar{p})\mu\nu(\alpha\mu - \lambda\mu p - \alpha\lambda - \lambda\mu)}\lambda^2,$$

$$\Pi_2'(1) = \frac{\lambda(\mu - \nu)(p + 1) + \mu\nu p + \alpha\lambda + \alpha\nu}{(\alpha + \lambda\bar{p})\mu\nu(\alpha\mu - \lambda\mu p - \alpha\lambda - \lambda\mu)}\alpha\lambda^2.$$

## 5  Numerical Results

To illustrate our analysis, we now present some numerical examples. In Figs. 1 and 2, we depict the mean and the variance of the number of customers in orbit, versus the arrival rate $\lambda$ for different values of the retrial rate $\nu$ as indicated. We fix $\alpha = 1$, $\mu = 0.1$ and $p = 1$ such that the length of the second phase is 10 times that of the first phase and customers always join the orbit. From these figures, we observe that both the mean and the variance of the number of customers in the orbit increase with the arrival rate as expected, with a vertical asymptote at $\lambda = (2\alpha^{-1} + \mu^{-1})^{-1} = 0.833$ in accordance with the stability condition. Furthermore, the mean orbit size and the variance of the orbit size are smaller if the retrial rate increases. The effect that more retrials give more opportunities to leave the orbit and start service dominates the effect that more retrials lead to more collisions as well.

For fixed $\lambda = 0.08$ and the same $\alpha$ and $\mu$ as above, Figs. 3, 4 and 5 show the distribution of the orbit size for different retrial rates: $\nu = 0.1$ in Fig. 3, $\nu = 1$ in Fig. 4 and $\nu = 10$ in Fig. 5. First note that the $\pi_{i,j}$ are parallel when $j$ is large which is in line with our observations on the asymptotic behavior, see Remark 1. A comparison of these figures further reveals that by an increase of the retrial rate, not only lower orbit sizes or more likely, but the distributions also decay faster for increasing orbit sizes.

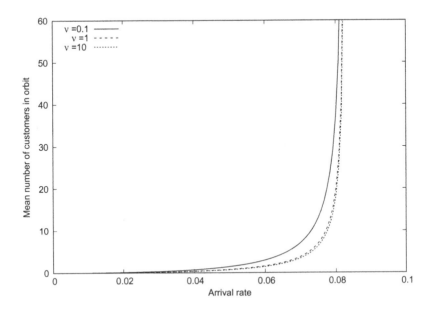

**Fig. 1.** Mean orbit size against $\lambda$.

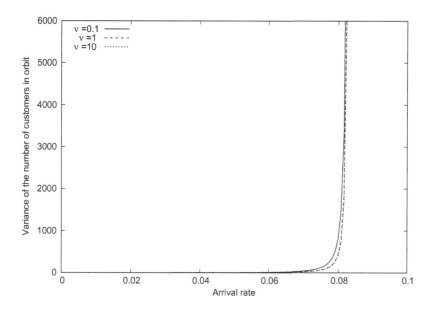

**Fig. 2.** Variance of orbit size against $\lambda$.

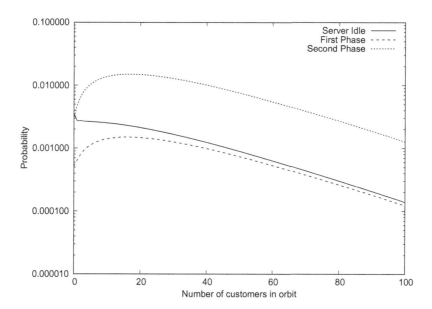

**Fig. 3.** Distribution of the number of customers in orbit $\lambda$ ($\nu = 0.1$).

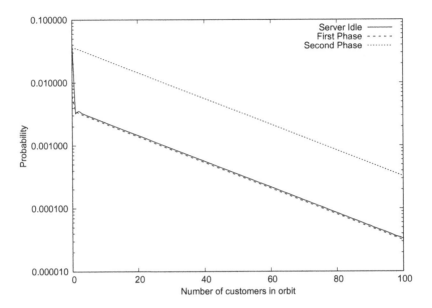

**Fig. 4.** Distribution of the number of customers in orbit $\lambda$ ($\nu = 1$).

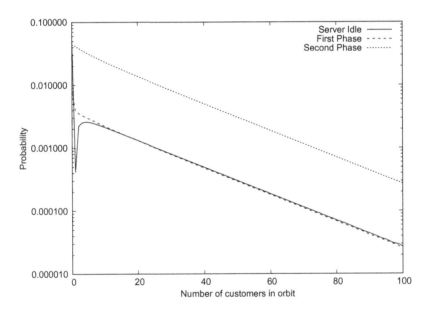

**Fig. 5.** Distribution of the number of customers in orbit $\lambda$ ($\nu = 10$).

# 6    Conclusion

In this paper, we considered a single server retrial queue with two phases of service. In the first phase, collisions may occur which sends the colliding customers to the orbit. On the other hand, in the second phase there is no collision. We have obtained explicit expressions for the generating functions of the stationary distributions of the orbit sizes when the server is in the different states. We also obtained the stability condition and recursive formulae for the joint stationary distribution which is convenient for numerical computation. We then extended the modeling assumptions by allowing for probabilistic abandonment by arriving customers.

For future work, the stability condition for the multi-server retrial queue with collisions will be considered. Furthermore, we found that the more general case with abandonment of retrying customers is not tractable by the present methodology. We therefore plan to investigate this extended queueing problem by a series expansion method. While the methodology is not exact, we hope to obtain good approximations of the joint stationary distribution and the corresponding moments.

# References

1. Khomichkov, I.I.: Computing the characteristics of a queueing system with repeated units and twin connections. Avtomatika i Telemekhanika **4**, 77–84 (1988)
2. Choi, B.D., Shin, Y.W., Ahn, W.C.: Retrial queues with collision arising from unslotted CSMA/CD protocol. Queueing Syst. **11**(4), 335–356 (1992). https://doi.org/10.1007/BF01163860
3. Kumar, B.K., Vijayalakshmi, G., Krishnamoorthy, A., Basha, S.S.: A single server feedback retrial queue with collisions. Comput. Oper. Res. **37**(7), 1247–1255 (2010)
4. Kumar, B.K., Rukmani, R., Thangaraj, V., Krieger, U.R.: A single server retrial queue with Bernoulli feedback and collisions. J. Stat. Theory Pract. **4**(2), 243–260 (2010). https://doi.org/10.1080/15598608.2010.10411984
5. Pakes, A.G.: Some conditions for ergodicity and recurrence of Markov chains. Oper. Res. **17**, 1058–1061 (1969)
6. Artalejo, J.R., Phung-Duc, T.: Single server retrial queues with two way communication. Appl. Math. Model. **37**(4), 1811–1822 (2013)

# Infinite-Server Bulk Queue with MMPP Arrivals

Anna Boyarkina[1] (ID), Svetlana Moiseeva[1] (ID), Michele Pagano[2] (ID),
Ekaterina Lisovskaya[1,3(✉)] (ID), and Alexander Moiseev[1] (ID)

[1] Tomsk State University, 36 Lenina Avenue, Tomsk 634050, Russian Federation
boyarkina.93@inbox.ru, smoiseeva@mail.ru, moiseev.tsu@gmail.com
[2] Department of Information Engineering, University of Pisa,
Via Caruso 16, 56122 Pisa, Italy
michele.pagano@iet.unipi.it
[3] Peoples' Friendship University of Russia (RUDN University),
6 Miklukho-Maklaya Street, Moscow 117198, Russian Federation
lisovskaya-eyu@rudn.ru

**Abstract.** We consider a queueing system with batch MMPP arrivals and an unlimited number of servers. An arriving group of customers occupies necessary number of free servers for a random time determined by a given probability distribution function. At the end of the service, the whole group releases all the servers at the same time. In the paper, we obtain the probability distribution of the number of customers in the system. To solve the problem, we use the dynamic screening method and asymptotic analysis such condition of high intensity of arrivals. It is shown that under condition the probability distribution of the number of customers in the system is Gaussian. The parameters of this Gaussian distribution are obtained in the paper.

**Keywords:** Infinite-server queue · Batch arrivals · Bulk service · Dynamic screening method · Asymptotic analysis

## 1 Introduction

Queueing theory is actively developing, however, there are still various real systems for which there are no suitable mathematical models describing the process of their evolution. Because of this, for example, difficulties arise when researching such objects that function according to a complex mechanism and whose states are difficult to predict. Recently, a scientific direction devoted to the study of queue networks with bulk arrival of customers and their bulk service has been intensively developing. Such models are effectively used as mathematical models of complex systems.

The publication has been prepared with the support of the "RUDN University Program 5-100" (recipient E. Lisovskaya, Visualization and Writing – review & editing).

M. Gribaudo et al. (Eds.): ASMTA 2019, LNCS 12023, pp. 158–170, 2020.
https://doi.org/10.1007/978-3-030-62885-7_12

The bulk (or batch) queueing systems are used to solve a wide class of scientific and applied problems. The variety of problems facing researchers led to the development of various modifications of such models: various types of arrival process and service disciplines, priorities, queues with losses, unreliable servers, etc. It is believed that the beginning of the QS studies with bulk services was found by N.T.J. Bailey [1] and F. Downton [4]. Rupert G. Miller Jr. in [9] considered models with batch arrivals and bulk service in which the size of groups was modeled as discrete random variables. For arbitrary distributions of intervals in arrival process and service times, the condition of ergodicity of the queue was found. Waiting time and busy period were obtained by the method of Markov chains for the case of the Poisson arrival process, and the characteristics of the stationary queue distribution were obtained for the case of exponential service. N.A. Jaiswal [7] proposed a queueing model in which the length of the bulk of arriving customers depends on the number of customers that are under service. To study the system, the author used the method of Markov chains too. P.D. Finch [5] studied a system with a finite queue and exponential service that is performed on bulks of fixed length. J.K. Goyal in [6] analyzed a system with hyper-exponential arrivals and exponential service of variable-length bulks using the method of generating functions. The results of these and many other works are presented in reviews [2] and [3]. Further, many authors turned to the study of systems with bulk service introducing a huge number of modifications. A review of the current state of the research problem was considered in [16,17]. Speaking about the evolution of research methods, it should be noted that most of the works use closely related methods based on Markov chains and generating functions. It is well known that their relative simplicity is leveled by the limited capabilities of analyzing the characteristics of complex systems. The method of supplementary variable is devoid of the indicated drawback, but the high complexity of its implementation for bulk queues causes its very rare application. To date, single-line queues with bulk servicing are sufficiently studied, while no general approach has been developed for multi-line and infinite-server non-Markov models.

This paper demonstrates the application of the dynamic screening method [11,13,14] to solve the problem of studying infinite-server queues with bulk service.

## 2    Mathematical Model

Let us consider an open queueing system with an unlimited number of servers, a bulk arrival process and bulk service (see Fig. 1). The peculiarity of this system is that an arrived group of customers leaves the system after its service in the same composition. Groups of customers arrive according to the Batch Markov Modulated Poisson Process (Batch MMPP) with an underlying Markov chain $k(t)$ given by the generator matrix $\mathbf{Q} = \|q_{\nu k}\|$, $\nu, k = 1, ..., K$. The conditional intensities of bulk arrivals are given by the diagonal matrix $\mathbf{\Lambda} = \mathrm{diag}\{\lambda_1, ..., \lambda_K\}$. The size of an arriving group is a discrete random variable $\xi$ given by probabilities $P\{\xi = i\} = p_i$, $i = 1, 2, ....$. Arrived batch of size $i$ occupies any $i$

free servers for a random time $\tau > 0$ with a probability distribution function $B(x) = P\{\tau < x\}$, and after the service completion all customers of the batch release servers occupied by them at the same time.

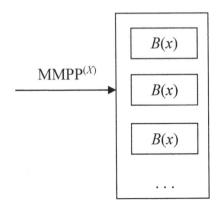

**Fig. 1.** Infinite-server bulk queue with MMPP arrivals

The goal of the study is to obtain the probability distribution of the number of customers in the system. The complexity of the analytical study is a prerequisite that all customers of a group leave the system at the same time, and therefore, there is a need to take into account the sizes of the received groups as well.

To solve the problem, it is proposed to use the dynamic screening method [10,11,13–15] as it was applied to resource systems [8,12]. This approach, in combination with asymptotic analysis, will further allow us to study more complex systems with correlated arrival processes.

## 3    Dynamic Screening Method

Let the system be empty at the initial time moment $t_0$. Denote the number of bulks that are under service in the system at moment $t \geq t_0$ by $K(t)$, and denote the total number of customers in the system by $V(t) = \sum_{l=1}^{K(t)} v_l$, where $v_l$, $l = 1, \ldots, K(t)$ are the sizes of the $l$-th bulk.

The goal is to find the stationary probability distribution of the two-dimensional random process $\{K(t), V(t)\}$. Due to the fact that the process is not Markovian, a modified method of dynamic screening is used. This method allows taking into account the resources occupied by customers who are in the system.

Let us fix a certain time moment $T > t_0$. We mark all epochs of bulks arrivals on axis 0 (Fig. 2). Further, we consider only those bulks that have not completed their service by the time moment $T$. We generate the points of the screened process (axis 1) from the moments of arrival of these bulks. Consider the probability that a group of size $v$ that arrived at the system at the moment $t$ will not finish its service until the moment $T$. Let us denote this probability by $S(t,T)$. It is obvious that $S(t,T) = 1-B(T-t)$. On the other hand, the group of customers will finish their service and free all the servers that they occupied when they entered the system with probability $1 - S(t,T)$. In the screened process, such groups are not taken into account.

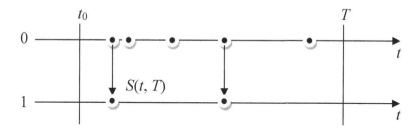

**Fig. 2.** Screening of arrived bulks of customers

Let us denote the number of screened group arrivals in the interval $[t_0, t)$ by $N(t)$, and the total number of screened customers in the same interval by $W(t)$. It is known [8] that at the time moment $t = T$, the probability distributions of random variables $\{K(T), V(T)\}$ and $\{N(T), W(T)\}$ coincide:

$$P\{K(T) = N, V(T) = W\} = P\{N(T) = N, W(T) = W\}. \tag{1}$$

We consider three-dimensional random process $\{k(t), N(t), W(t)\}$ which is Markovian. For its probability distribution $P\{k(t) = k, N(t) = N, W(t) = W\} = P(k, N, W, t)$, $k = \overline{1, K}$, $N, W = \overline{0, \infty}$, applying the formula of total probability, we can write down the equalities:

$$P(k, N, W, t + \Delta t) = P(k, N, W, t)(1 - \lambda_k \Delta t)(1 + q_{kk}\Delta t) + $$
$$P(k, N, W, t)\lambda_k \Delta t(1 - S(t)) + $$
$$\lambda_k \Delta t S(t) \sum_{i=1}^{W} P(k, N-1, W-i, t)p_i + \sum_{s \neq k} P(s, N, W, t)q_{sk}\Delta t + o(\Delta t)$$

and derive the system of balance (Kolmogorov) differential equations:

$$\frac{\partial P(k, N, W, t)}{\partial t} = \lambda_k S(t)\Big[\sum_{i=1}^{W} P(k, N-1, W-i, t)p_i - P(k, N, W, t)\Big] + $$
$$\sum_{s} P(s, N, W, t)q_{sk}. \tag{2}$$

with initial condition

$$P(k, N, W, t_0) = \begin{cases} R(k), & \text{if } N = W = 0, \\ 0, & \text{otherwise.} \end{cases}$$

Here $R(k)$ is the stationary state probability distribution of the underlying Markov chain and satisfies the following conditions:

$$\sum_{k=1}^{K} R(k)q_{vk} = 0, \quad v = \overline{1, K}, \quad \sum_{k=1}^{K} R(k) = 1.$$

Let us introduce the partial characteristic functions

$$h(k, x, y, t) = \sum_{N=0}^{\infty} e^{jxN} \sum_{W=0}^{\infty} e^{jyW} P(k, N, W, t).$$

Then from (2) we can write the following system of differential equations:

$$\frac{\partial h(k, x, y, t)}{\partial t} = \lambda_k S(t) h(k, x, y, t) \left[ e^{jx} V^*(y) - 1 \right] + \sum_{s} h(s, x, y, t) q_{sk}.$$

with initial condition

$$h(k, x, y, t_0) = R(k), \quad k = \overline{1, K}.$$

Here and below $V^*(y) = M\{e^{jy\xi}\} = \sum_{i=0}^{\infty} e^{jyi} p_i$.

Let us rewrite the problem in the matrix form:

$$\frac{\partial \mathbf{h}(x, y, t)}{\partial t} = \mathbf{h}(x, y, t) \left[ \mathbf{\Lambda} S(t)(e^{jx} V^*(y) - 1) + \mathbf{Q} \right] \tag{3}$$

with the initial condition

$$\mathbf{h}(x, y, t_0) = \mathbf{r}, \tag{4}$$

where

$$\mathbf{h}(x, y, t) = [h(1, x, y, t), h(2, x, y, t), \ldots, h(K, x, y, t)],$$

$$\mathbf{r} = [R(1), R(2), \ldots, R(K)].$$

$\mathbf{r}$ is the row vector of the stationary state probability distribution of the underlying Markov chain and satisfies the conditions

$$\begin{cases} \mathbf{rQ} = \mathbf{0}, \\ \mathbf{re} = 1 \end{cases} \tag{5}$$

(here $\mathbf{e}$ is a column vector whose all entries are equal to 1).

## 4   Asymptotic Analysis

We will seek the solution of Eq. (3) by the method of asymptotic analysis under the condition of increasing intensity of arrivals and frequent changes of the state of the underlying Markov chain. Let us denote $\mathbf{\Lambda} = N\tilde{\mathbf{\Lambda}}$ and $\mathbf{Q} = N\tilde{\mathbf{Q}}$, where $N$ is a high-intensity parameter (theoretically $N \to \infty$). Substituting this into (3), we obtain:

$$\frac{1}{N}\frac{\partial \mathbf{h}(x,y,t)}{\partial t} = \mathbf{h}(x,y,t)\left[\tilde{\mathbf{\Lambda}}S(t)\left(e^{jx}V^*(y) - 1\right) + \tilde{\mathbf{Q}}\right]. \tag{6}$$

**Theorem 1.** *Under the conditions of increasing intensity of arrivals and frequent changes of the state of the underlying Markov chain, the asymptotic characteristic function of the first order for the process* $\{k(t), N(t), W(t)\}$ *may be calculated as follows:*

$$\mathbf{h}(x,y,t) \approx \mathbf{r}\exp\left\{N\lambda\left[jx + jya_1\right]\int_{t_0}^{t}S(\tau)d\tau\right\},$$

*where* $a_1 = M\{\xi\} = \sum_{i=0}^{\infty} ip_i$ *is the average size of a bulk.*

*Proof.* Let us make the following substitutions in Eq. (6):

$$\varepsilon = \frac{1}{N}, \quad x = \varepsilon\tilde{x}, \quad y = \varepsilon\tilde{y}, \quad \mathbf{h}(x,y,t) = \mathbf{f}_1(\tilde{x},\tilde{y},t,\varepsilon). \tag{7}$$

We obtain

$$\varepsilon\frac{\partial \mathbf{f}_1(\tilde{x},\tilde{y},t,\varepsilon)}{\partial t} = \mathbf{f}_1(\tilde{x},\tilde{y},t,\varepsilon)\left[\tilde{\mathbf{\Lambda}}S(t)(e^{j\varepsilon\tilde{x}}V^*(\varepsilon\tilde{y}) - 1) + \tilde{\mathbf{Q}}\right]. \tag{8}$$

Let us find asymptotic solution $\mathbf{f}_1(\tilde{x},\tilde{y},t) = \lim_{\varepsilon \to 0}\mathbf{f}_1(\tilde{x},\tilde{y},t,\varepsilon)$ of Eq. (8) when $\varepsilon \to 0$.

*Step 1.* Let $\varepsilon \to 0$ in (7). We derive

$$\mathbf{f}_1(\tilde{x},\tilde{y},t)\tilde{\mathbf{Q}} = \mathbf{0}.$$

Comparing this equality with the first equation of system (5), we can draw a conclusion that $\mathbf{f}_1(\tilde{x},\tilde{y},t)$ can be written in the form

$$\mathbf{f}_1(\tilde{x},\tilde{y},t) = \mathbf{r}\Phi_1(\tilde{x},\tilde{y},t), \tag{9}$$

where $\Phi_1(\tilde{x},\tilde{y},t)$ is some scalar function that satisfies initial condition

$$\Phi_1(\tilde{x},\tilde{y},t_0) = 1 \tag{10}$$

(this follows from (4)).

*Step 2.* Multiplying (8) on vector $\mathbf{e}$ at the right hand, performing limit transition $\varepsilon \to 0$, and taking into account (5), we derive the following differential equation for function $\Phi_1(\tilde{x}, \tilde{y}, t)$:

$$\frac{\partial \Phi_1(\tilde{x}, \tilde{y}, t)}{\partial t} = \Phi_1(\tilde{x}, \tilde{y}, t) \left[ \lambda S(t) \left( j\tilde{x} + j\tilde{y} a_1 \right) \right].$$

Its solution under initial condition (10) is as follows:

$$\Phi_1(\tilde{x}, \tilde{y}, t) = \exp \left\{ \lambda (j\tilde{x} + j\tilde{y} a_1) \int_{t_0}^{t} S(\tau) d\tau \right\},$$

where $\lambda = \mathbf{r}\tilde{\mathbf{A}}\mathbf{e}$. Substituting this expression into (9), we derive

$$\mathbf{f}_1(\tilde{x}, \tilde{y}, t) = \mathbf{r} \exp \left\{ \lambda (j\tilde{x} + j\tilde{y} a_1) \int_{t_0}^{t} S(\tau) d\tau \right\}.$$

Taking into account substitutions (7), we can write down the following asymptotic (for $\varepsilon \to 0$) approximate expression:

$$\mathbf{h}(x, y, t) \approx \mathbf{r} \exp \left\{ N\lambda (jx + jy a_1) \int_{t_0}^{t} S(\tau) d\tau \right\}.$$

The theorem is proved.

**Theorem 2.** *Under the conditions of increasing intensity of arrivals and frequent changes of the state of the underlying Markov chain, the asymptotic characteristic function of the second order for the process $\{k(t), N(t), W(t)\}$ may be calculated as follows:*

$$
\begin{aligned}
\mathbf{h}(x, y, t) \approx \mathbf{r} \exp \Bigg\{ & N\lambda (jx + jy a_1) \int_{t_0}^{t} S(\tau) d\tau + \\
& \frac{(jx)^2}{2} \left[ N\lambda \int_{t_0}^{t} S(\tau) d\tau + N\kappa \int_{t_0}^{t} S^2(\tau) d\tau \right] + \\
& \frac{(jy)^2}{2} \left[ N\lambda a_2 \int_{t_0}^{t} S(\tau) d\tau + N a_1^2 \kappa \int_{t_0}^{t} S^2(\tau) d\tau \right] + \\
& jx jy \left[ N\lambda a_1 \int_{t_0}^{t} S(\tau) d\tau + N\kappa a_1 \int_{t_0}^{t} S^2(\tau) d\tau \right] \Bigg\},
\end{aligned}
$$

where $a_2 = M\{\xi^2\} = \sum\limits_{i=0}^{\infty} i^2 p_i$ is the second initial moment of random variable $\xi$ (size of a bulk), $\lambda = \mathbf{r}\tilde{\mathbf{\Lambda}}\mathbf{e}$, $\kappa = 2\mathbf{g}(\tilde{\mathbf{\Lambda}} - \lambda\mathbf{I})\mathbf{e}$, and row vector $\mathbf{g}$ satisfies the system of linear matrix equations

$$\begin{cases} \mathbf{g}\tilde{\mathbf{Q}} = \mathbf{r}(\lambda\mathbf{I} - \tilde{\mathbf{\Lambda}}), \\ \mathbf{g}\mathbf{e} = 0. \end{cases}$$

*Proof.* Let us introduce function $\mathbf{h}(x,y,t)$ in the form

$$\mathbf{h}(x,y,t) = \mathbf{h_2}(x,y,t)\exp\left\{N\lambda(jx + jya_1)\int\limits_{t_0}^{t} S(\tau)d\tau\right\}.$$

Substituting this into (6), we obtain

$$\frac{1}{N}\frac{\partial\mathbf{h_2}(x,y,t)}{\partial t} + \lambda(jx + jya_1)S(t)\mathbf{h_2}(x,y,t) = \mathbf{h_2}(x,y,t)\left[\tilde{\mathbf{\Lambda}}S(t)(e^{jx}V^*(y) - 1) + \tilde{\mathbf{Q}}\right]. \tag{11}$$

Let us make here the following substitutions:

$$\varepsilon^2 = \frac{1}{N}, \quad x = \varepsilon\tilde{x}, \quad y = \varepsilon\tilde{y}, \quad \mathbf{h_2}(x,y,t) = \mathbf{f_2}(\tilde{x},\tilde{y},t,\varepsilon). \tag{12}$$

Then equation (11) takes the form

$$\varepsilon^2\frac{\partial\mathbf{f_2}(\tilde{x},\tilde{y},t,\varepsilon)}{\partial t} + \mathbf{f_2}(\tilde{x},\tilde{y},t,\varepsilon)\lambda\left(j\varepsilon\tilde{x} + j\varepsilon\tilde{y}a_1\right)S(t) = \mathbf{f_2}(\tilde{x},\tilde{y},t,\varepsilon)\left[\tilde{\mathbf{\Lambda}}S(t)(e^{j\varepsilon\tilde{x}}V^*(\varepsilon\tilde{y})) + \tilde{\mathbf{Q}}\right] \tag{13}$$

with initial condition

$$\mathbf{f_2}(\tilde{x},\tilde{y},t_0,\varepsilon) = \mathbf{r}. \tag{14}$$

Let us find the asymptotic solution

$$\mathbf{f_2}(\tilde{x},\tilde{y},t) = \lim\limits_{\epsilon\to 0}\mathbf{f_2}(\tilde{x},\tilde{y},t,\varepsilon) \tag{15}$$

of problem (13)–(14) when $\varepsilon \to 0$.

*Step 1.* Let us make asymptotic transition $\varepsilon \to 0$ in problem (13)–(14). We obtain the following system of equations:

$$\begin{cases} \mathbf{f_2}(\tilde{x},\tilde{y},t)\tilde{\mathbf{Q}} = \mathbf{0}, \\ \mathbf{f_2}(\tilde{x},\tilde{y},t_0) = \mathbf{r}. \end{cases}$$

Comparing it with system (5), we can draw the conclusion that $\mathbf{f_2}(\tilde{x},\tilde{y},t)$ can be written in the form

$$\mathbf{f_2}(\tilde{x},\tilde{y},t) = \mathbf{r}\Phi_2(\tilde{x},\tilde{y},t), \tag{16}$$

where $\Phi_2(\tilde{x}, \tilde{y}, t)$ is some scalar function that satisfies the initial condition

$$\Phi_2(\tilde{x}, \tilde{y}, t_0) = 1. \tag{17}$$

*Step 2.* Taking into account (15) and (16), we can represent function $f_2(\tilde{x}, \tilde{y}, t_0, \varepsilon)$ in the form of expansion

$$\mathbf{f_2}(\tilde{x}, \tilde{y}, t) = \Phi_2(\tilde{x}, \tilde{y}, t)\left[\mathbf{r} + \mathbf{g}\left(j\varepsilon\tilde{x} + j\varepsilon\tilde{y}a_1\right)S(t)\right] + \mathbf{O}(\varepsilon^2), \tag{18}$$

where $\mathbf{g}$ is some row vector and $\mathbf{O}(\varepsilon^2)$ is a row vector that consists of infinitesimals of order $\varepsilon^2$.

Let us substitute (18) into (13) and make limit transition $\varepsilon \to 0$. We obtain the following matrix equation for vector $\mathbf{g}$:

$$\mathbf{g}\tilde{\mathbf{Q}} = \mathbf{r}(\lambda\mathbf{I} - \tilde{\boldsymbol{\Lambda}}),$$

where $\mathbf{I}$ is the identity matrix.

*Step 3.* Multiplying both sides of equation (13) by vector $\mathbf{e}$ at the right hand and denoting $\kappa = 2\mathbf{g}(\tilde{\boldsymbol{\Lambda}} - \lambda\mathbf{I})\mathbf{e}$, we derive the following differential equation for scalar function $\Phi_2(\tilde{x}, \tilde{y}, t)$:

$$\frac{\partial\Phi_2(\tilde{x}, \tilde{y}, t)}{\partial t} = \Phi_2(\tilde{x}, \tilde{y}, t)\left[\frac{(j\tilde{x})^2}{2}(\lambda S(t) + \kappa S^2(t)) + \right.$$

$$\left. \frac{(j\tilde{y})^2}{2}\left(\lambda a_2 S(t) + \kappa a_1^2 S^2(t)\right) + j\tilde{x}j\tilde{y}\left(\lambda a_1 S(t) + \kappa a_1 S^2(t)\right)\right],$$

where $a_2 = M\{\xi^2\} = \sum_{i=0}^{\infty} i^2 p_i$ is the second initial moment of random variable $\xi$ (size of a bulk). The solution of this equation under initial condition (17) is as follows:

$$\Phi_2(\tilde{x}, \tilde{y}, t) = \exp\left\{\frac{(j\tilde{x})^2}{2}\left(\lambda\int_{t_0}^{t} S(\tau)d\tau + \kappa\int_{t_0}^{t} S^2(\tau)d\tau\right) + \right.$$

$$\frac{(j\tilde{y})^2}{2}\left(\lambda a_2\int_{t_0}^{t} S(\tau)d\tau + \kappa a_1^2\int_{t_0}^{t} S^2(\tau)d\tau\right) +$$

$$\left. j\tilde{x}j\tilde{y}\left(\lambda a_1\int_{t_0}^{t} S(\tau)d\tau + \kappa a_1\int_{t_0}^{t} S^2(\tau)d\tau\right)\right\},$$

Substituting it into (16), we obtain

$$
\mathbf{f_2}(\tilde{x}, \tilde{y}, t) = \mathbf{r} \exp \left\{ \frac{(j\tilde{x})^2}{2} \left( \lambda \int_{t_0}^{t} S(\tau)d\tau + \kappa \int_{t_0}^{t} S^2(\tau)d\tau \right) + \right.
$$

$$
\frac{(j\tilde{y})^2}{2} \left( \lambda a_2 \int_{t_0}^{t} S(\tau)d\tau + \kappa a_1^2 \int_{t_0}^{t} S^2(\tau)d\tau \right) +
$$

$$
\left. j\tilde{x}j\tilde{y} \left( \lambda a_1 \int_{t_0}^{t} S(\tau)d\tau + \kappa a_1 \int_{t_0}^{t} S^2(\tau)d\tau \right) \right\}.
$$

Performing here substitutions inverse to (12), we can write the asymptotic expression for characteristic function $h(x, y, t)$:

$$
\mathbf{h}(x, y, t) \approx \mathbf{r} \exp \left\{ N\lambda(jx + jya_1) \int_{t_0}^{t} S(\tau)d\tau + \right.
$$

$$
\frac{(jx)^2}{2} \left( N\lambda \int_{t_0}^{t} S(\tau)d\tau + N\kappa \int_{t_0}^{t} S^2(\tau)d\tau \right) +
$$

$$
\frac{(jy)^2}{2} \left( N\lambda a_2 \int_{t_0}^{t} S(\tau)d\tau + N\kappa a_1^2 \int_{t_0}^{t} S^2(\tau)d\tau \right) +
$$

$$
\left. jxjy \left( N\lambda a_1 \int_{t_0}^{t} S(\tau)d\tau + N\kappa a_1 \int_{t_0}^{t} S^2(\tau)d\tau \right) \right\}.
$$

The theorem is proved.

Function $h(x, y, t) = \mathbf{h}(x, y, t)\mathbf{e}$ is the characteristic function of the two-dimensional process $\{N(t), W(t)\}$. Substituting $t = T$ to satisfy the main formula of the dynamic screening method (1) and $t_0 \to -\infty$ to achieve steady-state regime, we obtain the following approximate expression for the characteristic function of the two-dimensional process under study $\{K(t), V(t)\}$:

$$
h(x, y, t) \approx \exp \left\{ N\lambda(jx + jya_1)b + \frac{(jx)^2}{2}(N\lambda b + N\kappa\beta) + \right.
$$

$$
\left. \frac{(jy)^2}{2}(N\lambda a_2 b + N\kappa a_1^2 \beta) + jxjy(N\lambda a_1 b + N\kappa a_1 \beta) \right\},
$$

(19)

where

$$
b = \int_0^{\infty} (1 - B(\tau))d\tau, \quad \beta = \int_0^{\infty} (1 - B(\tau))^2 d\tau.
$$

Characteristic function of the total number of customers in the system in stationary regime can be obtained from (19) by substituting $x = 0$. So, we obtain

$$h(y) = h(x,y)|_{x=0} \approx \exp\left\{ jyN\lambda a_1 b + \frac{(jy)^2}{2}(N\lambda a_2 b + Na_1^2\kappa\beta) \right\}.$$

Hence the stationary probability distribution of the total number of customers in the system $MMPP^{(v)}/GI^{(v)}/\infty$ is asymptotically Gaussian with mean $a = N\lambda a_1 b$ and variance $\sigma^2 = N\lambda a_2 b + Na_1^2\kappa\beta$.

## 5   Simulation and Numerical Examples

Let the parameters of the MMPP arrival process be as follows:

$$\mathbf{\Lambda} = N \begin{bmatrix} 0.6 & 0 & 0 \\ 0 & 1.2 & 0 \\ 0 & 0 & 1.4 \end{bmatrix}, \quad \mathbf{Q} = N \begin{bmatrix} -6 & 3 & 3 \\ 4 & -8 & 4 \\ 5 & 5 & -10 \end{bmatrix}.$$

Here we have $\lambda = 1$, so, intensity of bulk arrivals is equal to $N$. Let service time have gamma distribution with parameters $\alpha = 1.5, \beta = 1.5$, so $b = 1$.

To determine the applicability area of the obtained asymptotic results we performed simulation experiments for various values of arrivals intensity (parameter $N$). To determine the accuracy of the approximation we use the Kolmogorov distance

$$\Delta = \max_i |F_{em}(i) - F_{as}(i)|,$$

where $F_{em}(i)$ and $F_{as}(i)$ are empirical and asymptotic distribution functions of the total number of customers in the system.

Table 1 shows the Kolmogorov distance between the simulation and asymptotic probability distributions of the number of customers in the system for various values of the parameter $N$. From the table, we can conclude that the accuracy of the approximation increases with increasing parameter $N$ (with increasing intensity of the incoming flow).

**Table 1.** Values of distances between asymptotic and empirical probability distributions of the number of occupied devices in the system

| $N$ | 1 | 5 | 10 | 20 | 50 | 70 | 100 | 200 | 500 |
|---|---|---|---|---|---|---|---|---|---|
| $\Delta$ | 0.182 | 0.056 | 0.033 | 0.022 | 0.014 | 0.011 | 0.009 | 0.005 | 0.002 |

If we suppose that the Kolmogorov distance less than 0.05 is enough for our purposes, we can conclude that the obtained asymptotic approximations are applicable when $N \geq 10$.

# 6   Conclusion

In this paper, we have studied a mathematical model of an infinite-server system with bulk MMPP arrivals and bulk servicing. Using the modified dynamic screening method and asymptotic analysis, we have obtained a Gaussian approximation for the probability distribution of the total number of customers in the system in steady-state regime. Using numerical analysis and simulations we have found the applicability area for the approximation.

The approach that is introduced in the paper allows studying non-Markovian infinite-server bulk queueing systems with various arrival processes and general service, as well as infinite-server bulk queueing systems and networks with more complex configuration.

# References

1. Bailey, N.T.J.: On Queueing Processes with Bulk Service. J. Roy. Stat. Soc.: Ser. B (Methodol.) **16**(1), 80–87 (1954). https://doi.org/10.1111/j.2517-6161.1954.tb00149.x
2. Chaudhry, M.L., Templeton, J.G.C.: A First Course in Bulk Queues. Wiley, New York (1983). https://books.google.ru/books?id=8WZRAAAAMAAJ
3. Cohen, J.W.: The Single Server Queue. North-Holland series in Applied Mathematics and Mechanics. North-Holland Pub. Co., Amsterdam (1969). https://books.google.ru/books?id=j2dRAAAAMAAJ
4. Downton, F.: Waiting time in bulk service queues. J. Roy. Stat. Soc.: Ser. B (Methodol.) **17**(2), 256–261 (1955). https://doi.org/10.1111/j.2517-6161.1955.tb00199.x
5. Finch, P.D.: On the transient behavior of a queueing system with bulk service and finite capacity. Ann. Math. Stat. **33**(3), 973–985 (1962). https://doi.org/10.1214/aoms/1177704465
6. Goyal, J.K.: Queues with hyper-poisson arrivals and bulk exponential service. Metrika **11**(1), 157–167 (1967). https://doi.org/10.1007/BF02613587
7. Jaiswal, N.K.: A bulk-service queueing problem with variable capacity. J. Roy. Stat. Soc.: Ser. B (Methodol.) **23**(1), 143–148 (1961). https://doi.org/10.1111/j.2517-6161.1961.tb00397.x
8. Lisovskaya, E.Y., Moiseev, A.N., Moiseeva, S.P., Pagano, M.: Modeling of mathematical processing of physics experimental data in the form of a non-markovian multi-resource queuing system. Russ. Phys. J. **61**(12), 2188–2196 (2019). https://doi.org/10.1007/s11182-019-01655-6
9. Miller Jr., R.G.: A contribution to the theory of bulk queues. J. Roy. Stat. Soc.: Ser. B (Methodol.) **21**(2), 320–337 (1959). https://doi.org/10.1111/j.2517-6161.1959.tb00340.x
10. Moiseev, A.N., Nazarov, A.A.: Infinite-server queueing systems and networks. NTL, Tomsk (2015). https://elibrary.ru/item.asp?id=30092762. (in Russian)
11. Moiseev, A.: Asymptotic analysis of the queueing network $SM - (GI/\infty)^K$. Commun. Comput. Inf. Sci. **564**, 73–84 (2015). https://doi.org/10.1007/978-3-319-25861-4_7

12. Moiseev, A., Moiseeva, S., Lisovskaya, E.: Infinite-server queueing tandem with MMPP arrivals and random capacity of customers. In: Proceedings of the 31st Conference on Modelling and Simulation, pp. 673–679 (2017). https://doi.org/10.7148/2017-0673

13. Moiseev, A., Nazarov, A.: Queueing network $MAP - (GI/\infty)^K$ with high-rate arrivals. Eur. J. Oper. Res. **254**(1), 161–168 (2016). https://doi.org/10.1016/j.ejor.2016.04.011

14. Moiseev, A., Nazarov, A.: Tandem of infinite-server queues with Markovian arrival process. Commun. Comput. Inf. Sci. **601**, 323–333 (2016). https://doi.org/10.1007/978-3-319-30843-2_34

15. Nazarov, A.A., Moiseeva, S.P.: Asymptotic Analysis Method in Queueing Theory. NTL, Tomsk (2006). https://elibrary.ru/item.asp?id=19588439. (in Russian)

16. Pershakov, N.V., Samuylov, K.E.: M/G/1 queues with batch service and its application to the stream control transmission protocol performance analysis. Part I. Discrete Continuous Models Appl. Comput. Sci. **1**, 34–44 (2009). http://journals.rudn.ru/miph/article/view/8465. (in Russian)

17. Pershakov, N.V., Samuylov, K.E.: M/G/1 queues with batch service and its application to the stream control transmission protocol performance analysis. Part II. Discrete Continuous Models Appl. Comput. Sci. **2**, 43–53 (2009). http://journals.rudn.ru/miph/article/view/8707. (in Russian)

# Dissipativity of the Quantum Measurement Model

Alexander V. Zorin[1], Leonid A. Sevastianov[1,2][(✉)],
and Nikolay P. Tretyakov[3]

[1] Department of Applied Probability and Informatics, Peoples' Friendship University
of Russia (RUDN University), 6 Miklukho-Maklaya Street, Moscow 117198, Russia
{zorin-av,sevastianov-la}@rudn.ru
[2] Joint Institute for Nuclear Research (Dubna),
Joliot-Curie, 6, Dubna, Moscow Region 141980, Russia
[3] Russian State Social University, Wilhelm Pieck Street, 4, Build.1,
Moscow 129226, Russia
trn11@rambler.ru

**Abstract.** The theory of quantum measurements is an extremely important part of quantum. The results of the quantum measurements theory are important for experimental study of quantum-mechanical objects and verification of the theoretical structure of quantum mechanics. Developed by Holevo and Helstrom the model of quantum measurements is the most rigorous and complete from the mathematical point of view. Alternatively, a very common approach to the theory of quantum measurements is generated by the theoretical formalization of experimental measurements. The theoretical construction formed in this way is called the operational model of quantum measurements. Their equivalence established in Ozawa's papers allowed us to describe the measured values of the quantum object using Weyl quantization rule, applied to the classical "measured quantities" obtained from the original classical quantities by convolution with Wigner function built from the quantum state functions of the probe. The result is the dissipative quantum model.

**Keywords:** Open quantum system · Quantum measurements · Quantum master equation · Dissipative quantum system

## 1   Introduction

From the very beginning of the construction of the quantum-mechanical model of the microworld, its creators were aware of the probabilistic nature of quantum mechanics. This was most clearly formulated in the classic treatise by von Neumann [1]. Its modern formulation is briefly presented in book [18].

The problem of describing the measurement procedure in quantum mechanics was apprehended by its founders at the very beginning of developing the theory

The publication has been prepared with the support of the "RUDN University Program 5-100".

M. Gribaudo et al. (Eds.): ASMTA 2019, LNCS 12023, pp. 171–185, 2020.
https://doi.org/10.1007/978-3-030-62885-7_13

basics. In the Soviet Union, D.I. Blokhintsev [2,3] and Ya.P. Terletsky [4] studied this problem most thoroughly. The results of their studies can be formulated as the statistical correspondence principle [5]. In its final form, this principle was implemented by V.V. Kuryshkin. The construction he proposed is known as quantum mechanics with non-negative quantum distribution function (QDF) [6–9].

L. Cohen revealed [10] that quantum mechanics with non-negative QDF generalizes the model of the operational approach to quantum measurement proposed by K. Wodkiewicz [11]. In compressed form, the modern description of the operational quantum measurement model is given in Appendix 1. As a result, it was found [12] that the Kuryshkin-Wodkiewicz model of quantum measurement implements the scheme of indirect quantum measurement proposed by Braginsky and Khalili [13,14]. A modern description of the indirect quantum measurement model is briefly presented in Appendix 2.

## 2    Basic Quantum Equations

In contrast to closed systems, the quantum dynamics of an open system generally cannot be presented in terms of unitary temporal evolution (see book [18]). The dynamics of an open quantum system is properly described by the density matrix evolution equation, the so-called master kinetic equation (see Appendix 3). Let us restrict ourselves to the simplest case of Markovian dynamics of open systems.

In analogy with the Kolmogorov-Chapman differential equation in the classical probability theory, the quantum dynamical equation is a linear differential equation of the first order for the reduced density matrix, known as the master quantum Markovian equation in the Lindblad form.

### 2.1    Markovian Quantum Master Equation

Let a quantum dynamical semigroup be given, then under certain conditions a generator of semigroups exists, i.e., such linear mapping $\hat{L}$ that the relation

$$\hat{V}(t) = \exp\left\{t\hat{L}\right\} \tag{1}$$

is valid. From this relation the differential equation of the first order in $t$ for the reduced density matrix follows:

$$\frac{d}{dt}\hat{\rho}_S(t) = \hat{L}\hat{\rho}_S(t), \tag{2}$$

referred to as the Markovian master quantum equation. The generator $\hat{L}$ is a superoperator acting in the ring $\{\hat{\rho}_S(t)\}$ of operators having a unit trace in $\mathcal{H}_S$. It is a generalization of Liouville superoperator (65), (66) for closed quantum systems.

Let us restrict ourselves to a finite-dimensional subspace $\{\varphi_k\} \subset \mathcal{H}_B$ with the dimension $N$ for density matrices in $\mathcal{H}_B$. Then in the space of matrices having the dimension $N^2$ let us choose a basis of $N$-dimensional matrices $\hat{F}_j$, orthonormalized with respect to the scalar product

$$\left(\hat{F}_i, \hat{F}_j\right) \equiv tr_S\left\{\hat{F}_i^\dagger \hat{F}\right\} = \delta_{ij}. \tag{3}$$

If we choose $\hat{F}_{N^2} = \left(^1/_N\right)^{1/2} \hat{I}_N$, then $tr_S\hat{F}_j = 0$ for $j = 1, 2, ..., N^2 - 1$. This basis of $N$-dimensional operators in $\mathcal{H}_N \in Span\{\varphi_k\}^N$ can be used to decompose acting in $\mathcal{H}_B$:

$$\hat{W}_{ij}(t) = \sum_{k=1}^{N^2} \hat{F}_k\left(\hat{F}_k, \hat{W}_{ij}(t)\right). \tag{4}$$

In this representation, the dynamical mapping $\hat{V}(t)$ is written in the form (if it does not go beyond the $B\left(\mathcal{H}_{N^2}\right)$ class of $N$-dimensional density matrices)

$$\hat{V}(t)\hat{\rho}_S = \sum_{i,j=1}^{N^2} c_{ij}(t) \hat{F}_i\hat{\rho}_S\hat{F}_j^\dagger, \tag{5}$$

where

$$c_{ij}(t) = \sum_{k,l}\left(\hat{F}_i, \hat{W}_{kl}(t)\right)\left(\hat{F}_j, \hat{W}_{kl}(t)\right)^* \tag{6}$$

is a positive Hermitian matrix.

From the decomposition (5) the explicit form of the generator $\hat{L}$ of the open quantum system follows:

$$\hat{L}\rho_S = -\frac{i}{\hbar}\left[H, \rho_S\right] + \sum_{i,j=1}^{N^2-1} a_{ij}\left(F_i\rho_S F_j^\dagger - \frac{1}{2}\left\{F_j^\dagger F_i, \rho_S\right\}\right), \tag{7}$$

where

$$a_{ij} = \lim_{t\to 0}\frac{c_{ij}(t)}{t}, a_{iN^2} = \lim_{t\to 0}\frac{c_{iN^2}(t)}{t}, a_{N^2N^2} = \lim_{t\to 0}\frac{c_{N^2N^2}(t) - N}{t} \tag{8}$$

and $\{*, *\}$ denotes an anticommutator.

Since the matrix of coefficients $a_{ij}$ is positive and Hermitian, it is diagonalised as

$$uau^\dagger = \begin{pmatrix} \gamma_1 & & \\ & \ddots & \\ & & \gamma_{N^2-1} \end{pmatrix} \tag{9}$$

with non-negative $\gamma_j$.

Let us introduce a new set of "basis" operators $A_k$

$$F_j = \sum_{k=1}^{N^2-1} u_{kj} A_k. \tag{10}$$

Then the first standard form (7) of the generator $\hat{L}$ transforms to the diagonal form

$$\hat{L}\rho_S = -\frac{i}{\hbar}[H, \rho_S] + \sum_{k=1}^{N^2-1} \gamma_k \left( A_k \rho_S A_k^\dagger - \frac{1}{2} A_k^\dagger A_k \rho_S - \frac{1}{2} \rho_S A_k^\dagger A_k \right). \tag{11}$$

The first term in the right-hand side of Eq. (11) (the most general form of the generator of a quantum dynamical semigroup) represents the unitary part of dynamics. The second term in Eq. (11) is called a dissipator; in this term $A_k$ are dimensionless and $\gamma_k$ have the dimension of inverse time. They depend on the environment correlation functions and play the role of relaxation rates for various types of damping in an open system.

Lindblad derived (3)–(11) in the general case of a separable Hilbert space $\mathcal{H}_S$ and operators $F$ and $A$ in $\mathcal{H}_S$.

The master quantum equation can be written in the form

$$\frac{d}{dt}\hat{\rho}_S(t) = -\frac{i}{\hbar}\left[\hat{H}(t), \hat{\rho}_S(t)\right] + D(\rho_S(t)). \tag{12}$$

where $D(\rho_S(t))$ is a dissipator having the form

$$D(\rho_S(t)) = \sum_{k=1}^{N^2-1} \gamma_k \left( A_k \rho_S A_k^\dagger - \frac{1}{2} A_k^\dagger A_k \rho_S - \frac{1}{2} \rho_S A_k^\dagger A_k \right) \tag{13}$$

Note that the generator $\hat{H}$ in Eq. (12) generally cannot be considered as a free Hamiltonian $\hat{H}_S$ of the reduced system $S$. The Hamiltonian $\hat{H}$ can incorporate additional terms generated by the interaction of $S$ and $B$.

Note also, that the generator $\hat{L}$ of the form (11) does not determine uniquely the form of the Hamiltonian $\hat{H}$ and the Lindblad operators $A_k$. The superoperator (11) is invariant with respect to the following transformations:

(i) the unitary transformations of the set of Lindblad operators

$$\sqrt{\gamma_k} A_k \rightarrow \sqrt{\gamma'_k} A'_k = \sum_j u_{kj} \sqrt{\gamma_j} A_j, \tag{14}$$

where $u_{kj}$ is a unitary matrix,

(ii) the inhomogeneous transformations

$$A_k \rightarrow A' = A_k + a_k$$
$$H \rightarrow H' = H + \frac{\hbar}{2i}\sum_{k=1}^{N^2-1} \gamma_k \left( a_k^* A_k - a_k A_k^\dagger \right) + b \tag{15}$$

where $a_j$ are complex numbers, $b$ is a real number. Thanks to (ii), one can use Eq. (15) to choose zero-trace operators $A_k$.

## 2.2 Adjoint Quantum Master Equation

The theory of quantum dynamical semigroups is commonly formulated in terms of the Heisenberg picture rather than Schrdinger one. The dynamical mapping in the Heisenberg picture is denoted by $\hat{V}^{\dagger}(t)$ and acts on the transformation operators $\hat{A}$ in the Hilbert space $\mathcal{H}_S$ of the open system, so that

$$tr_S \{A (V (t) \rho_S)\} = tr_S \left\{ (V^{\dagger} (t) A) \rho_S \right\} \tag{16}$$

for all $\rho_S$. In the Heisenberg picture, $V^{\dagger}(t)$ transforms positive operators into positive ones and preserves the unit operator

$$V^{\dagger}(t) \hat{I}_S = \hat{I}_S. \tag{17}$$

In complete analogy with the closed quantum systems, open quantum systems allow one to define for every observable operator $A$ (in the Schrdinger picture) the corresponding operator $\hat{A}(t)$ in the Heisenberg picture. This is implemented using the relation (16)

$$tr_S \{A (V (t,0) \rho_S)\} = tr_S \left\{ (V^{\dagger} (t,0) A) \rho_S \right\} = r_S \{A (t) \rho_S\} \tag{18}$$

In the definitions of $A(t)$ and $\hat{L}^{\dagger}$ we used the propagator

$$\hat{V}(t,t_0) = \underset{\leftarrow}{T} \exp \left\{ \int_{t_0}^{t} ds \hat{L}(s) \right\} \tag{19}$$

and the conjugate propagator

$$\hat{V}^{\dagger}(t,t_0) = \underset{\rightarrow}{T} \exp \left\{ \int_{t_0}^{t} ds \hat{L}^{\dagger}(s) \right\}, \tag{20}$$

where $\underset{\rightarrow}{T}$ is the anti-chronological time-ordering operator.

The conjugate operator $\hat{L}^{\dagger}$ is defined by the relation

$$tr_O \left\{ A \left( \hat{L}(t) \rho_O \right) \right\} = tr_O \left\{ \left( \hat{L}^{\dagger}(t) A \right) \rho_O \right\}, \tag{21}$$

which leads to the adjoint quantum master equation

$$\frac{d}{dt} A(t) = \hat{L}^{\dagger} A(t) \equiv \frac{i}{\hbar} [H, A(t)] +$$
$$+ \sum_k \gamma_k \left( A_k^{\dagger} A(t) A_k - \frac{1}{2} A(t) A_k^{\dagger} A_k - \frac{1}{2} A_k^{\dagger} A_k A(t) \right) \tag{22}$$

## 3    Kuryshkin-Wodkiewicz Quantum Measurement Model

The Kuryshkin-Wodkiewicz model of quantum measurement implements the scheme of indirect measurement, when a quantum object $O$ interacts with a quantum probe $P$. Thus, $S$ is a subsystem of the composite total system $O + P$, which is assumed closed and obeying the Hamilton dynamics. The evolution dynamics of $O$ depends on the internal dynamics and the interaction with the probe $P$.

Let us denote by $\mathcal{H}_O$ the Hilbert space of the object $O$, and by $\mathcal{H}_P$ the Hilbert space of the probe $P$. The Hilbert space of the total system $O + P$ is described by the tensor product $\mathcal{H} = \mathcal{H}_O \otimes \mathcal{H}_P$. The total Hamiltonian of the composite system has the form

$$\hat{H}(t) = \hat{H}_O \otimes \hat{I}_P + \hat{I}_O \otimes \hat{H}_P + \hat{H}_{\text{int}}(t) \tag{23}$$

The observables of the composite system related to $O$ have the form $\hat{A} \otimes \hat{I}_P$, where $\hat{A}$ is an operator in the space $\mathcal{H}_O$ and $\hat{I}_P$ is the unit operator in $\mathcal{H}_P$. If the state of the composite system is described by the density matrix $\rho$, then the mean values of the observables of the measured quantum object are written in the form

$$\left\langle \hat{A} \right\rangle_O = tr_O \left\{ \hat{A} \hat{\rho}_O \right\} \tag{24}$$

where

$$\hat{\rho}_O = tr_P \hat{\rho}$$

is the reduced density matrix of the quantum object $O$. Here $tr_O$ denotes a partial trace with respect to $\mathcal{H}_O$, and $tr_P$ a partial trace with respect to $\mathcal{H}_P$. The reduced density matrix $\hat{\rho}_O$ is the key quantity studied in the description of quantum measurements. From the definition presented in Appendix 2 it follows that

$$\hat{\rho}_O(t) = tr_P \left\{ \hat{U}(t, t_0) \hat{\rho}(t_0) \hat{U}^\dagger(t, t_0) \right\} \tag{25}$$

and

$$\frac{d}{dt} \hat{\rho}_O(t) = -\frac{i}{\hbar} tr_P \left[ \hat{H}(t), \hat{\rho}(t) \right]. \tag{26}$$

Generally, the dynamics of the measured quantum object, defined by the relations (25), (26), has extremely complex structure. We assume the dynamics to have no memory, so that it can be simplified to the structure of a quantum dynamical semigroup.

Let us at the initial moment of time $t_0 = 0$ prepare the state of the composite system $O + P$ as an uncorrelated product $\hat{\rho}(0) = \hat{\rho}_O(0) \otimes \hat{\rho}_P$, where $\hat{\rho}_O(0)$ is the initial state of the quantum object $O$, and $\hat{\rho}_P$ is the initial state of the quantum probe $P$.

The transformation of the measured quantum object from the initial moment of time $t_0 = 0$ to a certain moment $t_0 > 0$ can be written as

$$\hat{\rho}_O(0) \rightarrow \hat{\rho}_O(t) = \hat{V}(t) \hat{\rho}_O(0) \equiv tr_P \left\{ \hat{U}(t, 0) (\hat{\rho}_O(0) \otimes \hat{\rho}_P) \hat{U}^\dagger(t, 0) \right\} \tag{27}$$

If at the final moment of time $t$ the state of the probe $P$ coincides with the initial one, $\hat{\rho}_P$, then the relation (27) defines the mapping of the space of density matrices of the reduced system $S(\mathcal{H}_O)$ onto itself:

$$\hat{V}(t) : S(\mathcal{H}_O) \rightarrow S(\mathcal{H}_O) \tag{28}$$

The dynamical mapping can be fully characterized by operators acting in the Hilbert space $\mathcal{H}_O$ of the quantum object.

Let us use the representation of the probe density matrix in terms of the basis states $\{\varphi_k\} \subset \mathcal{H}_P$ :

$$\hat{\rho}_P = \sum a_j \, |\varphi_j\rangle \, \langle \varphi_j| \tag{29}$$

This relation reduces Eq. (27) to the form

$$\hat{V}(t)\,\hat{\rho}_O = \sum_{i,j} \hat{W}_{ij}(t)\,\hat{\rho}_O \hat{W}_{ij}^\dagger(t), \tag{30}$$

where $\hat{W}_{ij}(t)$ is an operator in $\mathcal{H}_O$, defined as

$$\hat{W}_{ij}(t) = \sqrt{a_j} \left\langle \varphi_i \left| \hat{U}(t,0) \right| \varphi_i \right\rangle. \tag{31}$$

From Eq. (30) it is seen that the dynamical mapping $\hat{V}(t)$ has the form of an operation that describes a generalized quantum measurement.

Let the classical instrument measure at the moment of time $\tau$ the probe observable

$$\hat{R} = \sum_m r_m \, |\varphi_m\rangle \, \langle \varphi_m| \tag{32}$$

with nondegenerate discrete spectrum. $\hat{R}$ is a self-adjoint operator in $\mathcal{H}$. The probability $P(m)$ of the outcome $r_m$ in this measurement is

$$P(m) = tr\,\{|\varphi_m\rangle \, \langle \varphi_m|\} = tr\left\{\hat{U}^\dagger \, |\varphi_m\rangle \, \langle \varphi_m| \, U\,(\rho_0 \otimes \rho_p)\right\}, \tag{33}$$

where the trace is calculated over the space of the composite system. The partial trace over the spaces $\mathcal{H}_0$ and $\mathcal{H}$ are denoted as $tr_0$ and $tr$. In this case the expression for $P(m)$ takes the form

$$P(m) = tr_0\,\{F_m\rho_0\}, \tag{34}$$

where the effect $F_m$ has the form

$$F_m\rho_0 = tr\left\{\hat{U}^\dagger \, |\varphi_m\rangle \, \langle \varphi_m| \, U\,(\rho_0 \otimes \rho_p)\right\}. \tag{35}$$

$F_m$ is a positive operator in $\mathcal{H}_0$ that satisfies the condition

$$\sum_m F_m\rho_0 = \sum tr\left\{\hat{U}^\dagger \, |\varphi_m\rangle \, \langle \varphi_m| \, U\,(\rho_0 \otimes \rho_p)\right\} = tr\,\{\rho_0 \otimes \rho_p\} = \rho_0. \tag{36}$$

The indirect measurement operation is obtained by applying the von Neumann-Luders projection postulate to the measurement of the probe, which determines the following state of the quantum object:

$$\rho'_m = P(m)^{-1} \, \langle \varphi_m| \, U\,(\rho_0 \otimes \rho_p)\, \hat{U}^\dagger \, |\varphi_m\rangle. \tag{37}$$

If the density matrix of the probe has the form

$$\rho = \sum_k a_k \, |\varphi_k\rangle \, \langle\varphi_k|,$$

(38)

then the operation on the quantum object has the form

$$\Phi_m(\rho_0) = \sum_k \Omega_{mk}\rho_0\Omega_{mk}^\dagger,$$

(39)

where

$$\Omega_{mk}(t) = \sqrt{a_k} \, \langle\varphi_m| \, U(t) \, |\varphi_m\rangle \, .$$

(40)

When using these operators, the effect has the form

$$F_m = \sum_k \Omega_{mk}^\dagger \Omega_{mk}.$$

(41)

Equation (30) reduces to the form

$$\hat{V}(t)\,\hat{\rho}_O = \sum_k \hat{W}_{mk}(t)\,\hat{\rho}_O\hat{W}_{mk}^\dagger(t),$$

(42)

A generator of the quantum dynamical semigroup is a linear mapping $\hat{L}$, such that

$$\hat{V}(t) = \exp\left\{t\hat{L}\right\}$$

(43)

From this equation the differential equation of the first order in $t$ for the reduced density matrix follows

$$\frac{d}{dt}\hat{\rho}_O(t) = \hat{L}\hat{\rho}_O(t),$$

(44)

referred to as the Markovian quantum master equation.

The quantum master equation can be written as

$$\frac{d}{dt}\hat{\rho}_O(t) = -\frac{i}{\hbar}\left[\hat{H}_O(t), \hat{\rho}_O(t)\right] + D(\rho_O(t))$$

(45)

where the dissipator $D(\rho_O(t))$ (the operator that determines the rate of dissipation [19,20]) has the form:

$$D(\rho_O(t)) = \sum_k \left\{ \frac{d\hat{W}_{mk}(t)}{dt}\hat{\rho}_O\hat{W}_{mk}^\dagger(t) + \hat{W}_{mk}(t)\hat{\rho}_O\frac{d\hat{W}_{mk}^\dagger(t)}{dt} \right\}$$

(46)

The conjugate operator $\hat{L}^\dagger$ is defined by the relation

$$tr_O\left\{A\left(\hat{L}(t)\,\rho_O\right)\right\} = tr_O\left\{\left(\hat{L}^\dagger(t)\,A\right)\rho_O\right\}$$

(47)

and determines the adjoint quantum master equation

$$\frac{d}{dt} A(t) = \hat{L}^\dagger A(t) \equiv \frac{i}{\hbar} [H_O, A(t)] + \sum_k \left\{ \frac{d\hat{W}_{mk}(t)}{dt} \hat{\rho}_O \hat{W}_{mk}^\dagger(t) + \hat{W}_{mk}(t) \hat{\rho}_O \frac{d\hat{W}_{mk}^\dagger(t)}{dt} \right\}$$
(48)

The presence of a nonzero dissipator in the Liouville operator (in the Lindblad form of the quantum kinetic equation) indicates the fact that the Kuryshkin-Wodkiewicz model of quantum measurement is a dissipative quantum system.

In Refs. [15] the authors used the technique by V.E. Tarasov [16] for quantization of dissipative classical systems to show that the Kuryshkin-Wodkiewicz rule of quantizing a classical observable $A(q, p)$ is equivalent to the Weyl quantization rule for a "dissipative" classical observable $A * W_\rho(q, p)$. In this case $W_\rho(q, p)$ is the Wigner distribution function of the mixed state matrix $\hat{\rho}_P = \sum a_j |\varphi_j\rangle \langle\varphi_j|$ of the probe, the quantum part of the measuring instrument.

In the papers by Tarasov [16] it is shown that according to the Sedov's scheme $D(\rho_O(t))$ can be presented as $\left[\hat{A}_H, \hat{W}\right]$, where $\hat{W}$ is a non-Hamiltonian operator. Therefore, it is not surprising that our representation (48) contains the contributions of exactly this form (see details in the papers [15]).

# 4   Conclusion

Usually, the dynamics of an open quantum system are described in terms of the reduced density operator, which is obtained from the density operator of the entire system by averaging over environmental variables. To simplify the resulting dependence, various approximations are needed, which lead to a closed equation of motion for the density matrix of an open system. The most famous of them is the Markov approximation, which leads to a quantum basic equation, which, in turn, generates a quantum dynamic semigroup in the space of reduced density matrices.

To extract information about a quantum system, a measurement must be made. Consider a situation where an open system is measured by indirectly measuring its environment. Then an open system is a measurable quantum object, and the environment plays the role of a quantum probe. The latter is measured using a classical instrument, as soon as correlations are formed between the object and the probe as a result of their interaction.

The model of quantum measurements eventually becomes a dissipative quantum system. The study of dissipative systems in quantum theory is of great theoretical interest and is important for practical applications. In this regard, the open press proposed various approaches to the formulation of dynamics in dissipative quantum systems. Two of them are considered in our work, and their equivalence is shown [15,17].

# Appendix 1. Operational Model of Quantum Measurement

The generalized theory of quantum measurements [Kraus 1983; Davies 1976; Bragiusky, Khalili 1992] is a natural generalization of the von Neumann-Luders projection postulate based on the generalized scheme of measurement (Fig. 1).

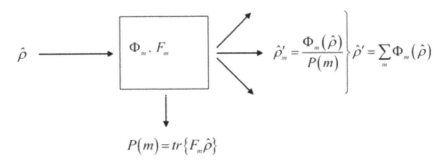

**Fig. 1.** Illustration of the generalized measurement scheme with operations $\Phi_m$ and effects $F_m$.

i) The measurement result $m$ is a classical random quantity with the probability distribution

$$P(m) = tr\{F_m\hat{\rho}\}, \tag{49}$$

where $F_m$ is a positive operator called an effect and satisfying the normalization condition

$$\sum_{m \in M} F_m = \hat{I}, \tag{50}$$

such that $P(m)$ is also normalized, i.e.

$$\sum_{m \in M} P(m) = 1. \tag{51}$$

ii) In the case of a selective measurement, the subensemble of those systems, for which the outcome $m$ was obtained, is described by the density matrix

$$\hat{\rho'}_m = \frac{\Phi_m(\hat{\rho})}{P(m)}, \tag{52}$$

where $\Phi_m = \Phi_m(\hat{\rho})$ is a positive superoperator called an operation, which maps positive operators onto positive operators. The operation $\Phi_m$ satisfies the condition of (statistical) consistency

$$tr\Phi_m(\hat{\rho}) = tr\{F_m\hat{\rho}\}, \tag{53}$$

which together with Eq. (49) defines the normalized density matrix, i.e.

$$tr\hat{\rho'}_m = \frac{tr\Phi_m(\hat{\rho})}{P(m)} = 1. \tag{54}$$

iii) For nonselective measurement, the density matrix takes the form

$$\hat{\rho}' = \sum_{m \in M} P(m) \hat{\rho}'_m = \sum_{m \in M} \Phi_m(\hat{\rho}), \tag{55}$$

which is normalized according to Eqs. 50 and 53:

$$tr\hat{\rho}' = \sum_{m \in M} tr\Phi_m(\hat{\rho}) = \sum_{m \in M} tr\{F_m\hat{\rho}\} = tr\hat{\rho} = 1. \tag{56}$$

The von Neumann-Luders scheme of ideal measurement is a special case of the generalized measurement scheme described above. Namely, let $\Delta \hat{E}_\alpha$ be an orthogonal decomposition of unit operator, then the set $M$ is a set of the intervals $\Delta r_\alpha$ of the decomposition $\Delta \hat{E}_\alpha$. For the result of measurement $= \Delta r_\alpha$ the effect is $F_m = \Delta \hat{E}_\alpha$ and the operation is the mapping $\Phi_m : \hat{\rho} \to \Phi_m(\rho) = \Delta \hat{E}_\alpha \rho \Delta \hat{E}_\alpha$.

As a natural generalization of the orthogonal decomposition, consider the operation

$$\Phi_m(\rho) = \Omega_m \rho \Omega_m^\dagger \tag{57}$$

and the corresponding effect

$$F_m = \Omega_m^\dagger \Omega_m, \tag{58}$$

where $\Omega$ are linear operators in the base Hilbert space, satisfying the normalization condition

$$\sum_{m \in M} F_m = \sum_{m \in M} \Omega_m^\dagger \Omega_m = \hat{I}. \tag{59}$$

## Appendix 2. Indirect Quantum Measurements

The scheme of indirect quantum measurement proposed in Ref. [Brag, Khal, 1992] is another option for generalized quantum measurement.

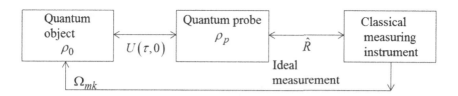

**Fig. 2.** Indirect measurement scheme.

The first element, the quantum object, i.e, a system from which the information is extracted, acts in $H_0$. The second element, the quantum probe, acts in $H_p$. Before the interaction with the quantum object, the quantum probe is prepared in an appropriate state with the density matrix $\rho_p$. Due to the subsequent interaction, the correlation between $\hat{\rho}_0$ and $\hat{\rho}_p$ is established.

The third element, the classical instrument, executes a measurement on the quantum probe when the object-probe interaction is over. The information about the object is extracted from the object-probe correlation.

An ideal indirect measurement satisfies three conditions.

i) Before the beginning of interaction the probe is in a well-prepared definite state $\hat{\rho}_p$; at this time the quantum object is in a certain state $\hat{\rho}_0$.
ii) The interaction begins after the preparation but before the measurement using the classical instrument, i.e., from $t = 0$ to $t = \tau > 0$.
iii) The classical instrument executes an ideal measurement on the probe in correspondence with the von Neumann-Luders projection postulate.

At the time moment $t = 0$ the total quantum system consisting of the object and the probe is described by the density matrix $\hat{\rho}_0 \otimes \hat{\rho}_p$ in the space $H = H_0 \otimes H_p$ and the Hamiltonian

$$H(t) = H_0 + H_p + H_I(t), \tag{60}$$

where $H_0$ and $H_p$ describe isolated evolutions of the object and the probe, and $H_I(t)$ describes their interaction that turns into zero at $t > \tau_0$.

The corresponding unitary operator $U$ has the form

$$U = \hat{U}(\tau, 0) = \underset{\leftarrow}{T} \exp\left\{-\frac{i}{\hbar} \int_0^\tau dt \hat{H}(t)\right\} \tag{61}$$

The initial density matrix $\rho(0) = \hat{\rho}_0 \otimes \hat{\rho}_p$ of the total system after the time $\tau$ turns into

$$\rho(\tau) = \hat{U}(\hat{\rho}_0 \otimes \hat{\rho}_p) \hat{U}^\dagger \tag{62}$$

Let the classical instrument measure the probe variable

$$\hat{R} = \sum_m r_m |\varphi_m\rangle \langle \varphi_m| \tag{63}$$

having nondegenerate discrete spectrum at the time moment $\tau$. $\hat{R}$ is a self-adjoint operator in $H_p$. The probability $P(m)$ of the outcome $r_m$ in this measurement is equal to

$$P(m) = tr\left\{|\varphi_m\rangle \langle \varphi_m|\right\} = tr\left\{\hat{U}^\dagger |\varphi_m\rangle \langle \varphi_m| U(\rho_0 \otimes \rho_p)\right\}, \tag{64}$$

where the trace is calculated over the space of the composite system. The partial traces over the spaces $H_0$ and $H_p$. are denoted by $tr_0$ and $tr_p$. In this case the expression for $P(m)$ takes the form

$$P(m) = tr_0\left\{F_m \rho_0\right\}, \tag{65}$$

where the effect $F_m$ has the form

$$F_m \rho_0 = tr\left\{\hat{U}^\dagger |\varphi_m\rangle \langle \varphi_m| U(\rho_0 \otimes \rho_p)\right\}. \tag{66}$$

$F_m$ is a positive operator in $\mathcal{H}_0$ satisfying the condition

$$\sum_m F_m \rho_0 = \sum_m tr \left\{ \hat{U}^\dagger \, |\varphi_m\rangle \langle \varphi_m| \, U \left( \rho_0 \otimes \rho_p \right) \right\} = tr \left\{ \rho_0 \otimes \rho_p \right\} = \rho_0. \quad (67)$$

The operation for the indirect measurement is obtained by applying the von Neumann-Luders projection postulate to the measurement of the probe, which specifies the following state of the quantum object

$$\rho'_m = P\left(m\right)^{-1} \langle \varphi_m | \, U \left( \rho_0 \otimes \rho_p \right) \hat{U}^\dagger \, |\varphi_m\rangle. \quad (68)$$

If the probe density matrix has the form

$$\rho_p = \sum_k p_k \, |\varphi_k\rangle \langle \varphi_k|, \quad (69)$$

then the operation on the quantum object has the form

$$\Phi_m\left(\rho_0\right) = \sum_k \Omega_{mk} \rho_0 \Omega^\dagger_{mk}, \quad (70)$$

where

$$\Omega_{mk} = \sqrt{p_k} \, \langle \varphi_m | \, U \, |\varphi_m\rangle. \quad (71)$$

When using these operators, the effect has the form

$$F_m = \sum_k \Omega^\dagger_{mk} \Omega_{mk}. \quad (72)$$

Thus, the operation and the effect have the form from the Kraus representation theorem (see book [18]).

If initially the probe was in a pure state, $\rho = |\varphi_k\rangle \langle \varphi_k|$, then we get the quantity

$$\Omega_m = \langle \varphi_m | \, \hat{U} \, |\varphi\rangle \quad (73)$$

proportional to the amplitude of the probe transition from the state $|\varphi\rangle$ to the eigenstate $\langle \varphi_m|$ of the measured observable $\hat{R}$. In this case $\Omega_m$ is an operator in $H_0$, describing the change in the object state caused by the outcome $r_m$ of the measurement on the probe. Additional summation over $k$ in the expression 70 appears due to the additional uncertainty in the case, when the initial state of the probe is a statistical mixture.

## Appendix 3. Dynamics of Open Systems

An open system is a quantum system $S$ that interacts with another quantum system $B$, referred to as environment. Thus, $S$ is a subsystem of a composite total system $S + B$, which is assumed closed and obeying Hamiltonian dynamics. The evolution dynamics of $S$ depends on its internal dynamics and the interaction with environment $B$.

This interaction leads to the system-environment correlation, therefore, the evolution of $S$ cannot be described in terms of unitary Hamiltonian dynamics. The dynamics of $S$ caused by the dynamics of the composite system is referred to as the reduced system dynamics and the system $S$ itself is called a reduced system.

Let us denote by $\mathcal{H}_S$ the Hilbert space of the system $S$ and by $\mathcal{H}_B$ the Hilbert space of the environment $B$. The Hilbert space of the total system $S + B$ is described by the tensor product $\mathcal{H} = \mathcal{H}_S \otimes \mathcal{H}_B$. The total Hamiltonian of the composite system has the form

$$\hat{H}(t) = \hat{H}_S \otimes \hat{I}_B + \hat{I}_S \otimes \hat{H}_S + \hat{H}_{\text{int}}(t) \tag{74}$$

The open system $S$ is picked out from the composite system $S + B$ using the criterion (principle) that all changes interesting for a researcher occur in $S$. The observables of the composite system related to $S$ have the form $\hat{A} \otimes \hat{I}_B$, where $\hat{A}$ is an operator in the space $\mathcal{H}_S$ and $\hat{I}_B$ is a unit operator in $\mathcal{H}_B$. If the state of the composite system is described by the density matrix $\rho$, then the mean values of the observables of the open system are calculated as

$$\left\langle \hat{A} \right\rangle_S = tr_S \left\{ \hat{A} \hat{\rho}_S \right\} \tag{75}$$

where

$$\hat{\rho}_S = tr_B \hat{\rho}$$

is the reduced density matrix of the open system $S$. Here $tr_S$ denotes the partial trace with respect to the space $H_S$ and $tr_B$ is the partial trace with respect to the space $H_B$. The reduced density matrix $\hat{\rho}_S$ is a key subject of study in the description of an open system. From the above definitions, it follows that

$$\hat{\rho}_S(t) = tr_B \left\{ \hat{U}(t, t_0) \hat{\rho}(t_0) \hat{U}^\dagger(t, t_0) \right\} \tag{76}$$

and

$$\frac{d}{dt} \hat{\rho}_S(t) = -\frac{i}{\hbar} tr_B \left[ \hat{H}(t), \hat{\rho}(t) \right]. \tag{77}$$

Application materials are taken from the book [18].

# References

1. Von Neumann, J.: Mathematical Foundations of Quantum Mechanics. Princeton University Press, Princeton (1955)
2. Blokhintzev, D.I.: The gibbs quantum ensemble and its connection with the classical ensemble. J. Phys. **II**(1), 71–74 (1940)
3. Blokhintzev, D., Nemirovsky, P.: Connection of the quantum ensemble with the gibbs classical ensemble. II. J. Phys. **III**(3), 191–194 (1940)
4. Terletsky, Y.P.: The limiting transition from quantum to classical mechanics. J. Exp. Theor. Phys. **7**(11), 1290–1298 (1937)

5. Zorin, A.V., Sevastianov, L.A., Belomestny, G.A.: Numerical search for the states with minimal dispersion in quantum mechanics with non–negative quantum distribution function. In: Li, Z., Vulkov, L., Waśniewski, J. (eds.) NAA 2004. LNCS, vol. 3401, pp. 613–620. Springer, Heidelberg (2005). https://doi.org/10.1007/978-3-540-31852-1_75

6. Kuryshkin, V.V.: La mecanique quantique avec une fonction nonnegative de distribution dans l'cespace des phases. Annales Inst. Henri Poincare **17**(1), 81–95 (1972)

7. Kuryshkin, V.V.: Une generalisation possible de la m ecanique quantique non relativiste. Compt. Rend. Acad. Sc. Paris. Serie B. **274**, 1107–1110 (1972)

8. Kuryshkin, V.V.: Some problems of quantum mechanics possessing a non-negative phase-space distribution function. Int. J. Theor. Phys. **7**(6), 451–466 (1973)

9. Zorin, A.V., Kuryshkin, V.V., Sevastyanov, L.A.: Description of the spectrum of a hydrogen-like atom. Bull. PFUR. Ser. Phys. **6**(1), 62–66 (1998)

10. Cohen, L.: Kuryshkin distributions and their generalizations. Discussion Questions of Quantum Physics, pp. 49–58. RUDN University (1993)

11. Wodkiewicz, K.: Operational approach to phase-space measurements in quantum mechanics. Phys. Rev. Lett. **52**, 1064 (1984)

12. Zorin, A.V.: The operational model of quantum measurement of Kuryshkin-Wodkiewicz. Discrete Continuous Models Appl. Comput. Sci. **2**, 43–55 (2012)

13. Braginsky, V.B., Vorontsov, Y.I., Halily, F.Y.: Optimal quantum measurements in detectors of gravitational radiation. Lett. J. Exp. Theor. Phys. **27**, 296–301 (1978)

14. Braginsky, V.B., Vorontsov, Y.I., Halily, F.Y.: Quantum features of the ponderomotive meter of electromagnetic energy. J. Exp. Theor. Phys. **73**, 1340–1343 (1977)

15. Sevastyanov, L., Zorin, A., Gorbachev, A.: Pseudo-differential operators in an operational model of the quantum measurement of observables. In: Adam, G., Buša, J., Hnatič, M. (eds.) MMCP 2011. LNCS, vol. 7125, pp. 174–181. Springer, Heidelberg (2012). https://doi.org/10.1007/978-3-642-28212-6_17

16. Tarasov, V.E.: Quantum dissipative systems. III. Definition and algebraic structure. Theor. Math. Phys. **110**(1), 57–67 (1997)

17. Breuer, H.-P., Petruccione, F.: Concepts and methods in the theory of open quantum systems. In: Benatti, F., Floreanini, R. (eds.) Irreversible Quantum Dynamics. Lecture Notes in Physics, vol. 622, pp. 65–79. Springer, Berlin (2003). https://doi.org/10.1007/3-540-44874-8_4

18. Breuer, H.P., Petruccione, F.: The Theory of Open Quantum Systems. Oxford University Press, Oxford (2007)

19. Holevo, A.S.: Probabilistic and Statistical Aspects of Quantum Theory. North-Holland, Amsterdam (1982)

20. Holevo, A.S.: Statistical Structure of Quantum Theory. Lecture Notes in Physics Monographs, vol. 67. Springer, Berlin (2001). https://doi.org/10.1007/3-540-44998-1

# Author Index

Printed in the United States
By Bookmasters